Revit Structure 2020 for Novices

(Learn By Doing)

CADSoft Technologies

Revit Strucrure 2020 for Novices:
Learn By Doing

CADSoft Technologies

ISBN 978-1-64057-082-5

DEDICATION

To students, who are dedicated to learning new technologies

THANKS

To the faculty and employees of CADSoft Technologies for their cooperation

Online Training Program Offered by CADSoft Technologies

CADSoft Technologies provides effective and affordable virtual online training on various software packages including Computer Aided Design, Manufacturing and Engineering (CAD/CAM/CAE), computer programming languages, animation, architecture, and GIS. The training is delivered 'live' via Internet at any time, any place, and at any pace to individuals as well as the students of colleges, universities, and CAD/CAM/CAE training centers. The main features of this program are:

Training for Students and Companies in a Classroom Setting

A team of highly experienced instructors and qualified engineers at CADSoft Technologies conduct the classes in classroom environment at its centers. This team has authored several textbooks that are rated "one of the best" in their categories and are used in various colleges, universities, and training centers in North America, Europe, and in other parts of the world.

Training for Individuals

CADSoft Technologies with its cost effective and time saving initiative strives to deliver the training in the comfort of your home or work place, thereby relieving you from the hassles of traveling to training centers.

Training Offered on Software Packages

CADSoft Technologies provides basic and advanced training on the following software packages:

***CAD/CAM/CAE**: CATIA, Creo Parameteric, SOLIDWORS, Autodesk Inventor, Solid Edge, NX, AutoCAD, AutoCAD LT, Customizing AutoCAD, AutoCAD Electrical, and ANSYS*

***Architecture and GIS**: Autodesk Revit (Architecture/Structure/MEP), AutoCAD Civil 3D, AutoCAD Map 3D, Autodesk Navisworks, Bentley Staad.Pro, Oracle Primavera P6, MS Project, ArcGIS, and Raster Design.*

***Animation and Styling**: Autodesk 3ds Max, 3ds Max Design, Autodesk Maya, Autodesk Alias, Pixologic ZBrush, and CINEMA 4D*

***Computer Programming**: C++, VB.NET, Oracle, AJAX, and Java*

For more information, please visit the following link:
https://www.cadsofttech.com

Table of Contents

Chapter 3: Setting up a Structural Project

Chapter 4: Structural Columns and Walls

Chapter 5: Foundations, Beams, Floors, and Open Web Joists

Chapter 6: Editing Tools

Chapter 7: Documenting Models and Creating Families

This page is intentionally left blank

Preface

Autodesk Revit 2020 for Structure

Autodesk Revit is a Building Information Modeling software developed by Autodesk. The structural tools in Revit are primarily used in structural engineering and they help the users to design structural models using both parametric 3D modeling and 2D drafting elements. The Revit building information model comprises of a physical representation of a building which is fully associated with analytical representation. The building information model in Revit can be used for structural designing, production of drawings, coordination of the project, and also for third party structural analysis application.

The **Revit Structure 2020 for Novices: Learn By Doing** textbook introduces the users to the spectacular realm of one of the most powerful software in the architectural quiver. This textbook is a gateway to power, skill, and competence in the field of architectural and interior presentations, drawings, and documentations.

This textbook is specially meant for professionals and students of architecture and interior design, facilities planners, and CAD professionals who are associated with the building construction and allied fields in the construction industry.

Special emphasis has been laid to explain new concepts, procedures, and methods in Revit by using sufficient text and graphical examples.The accompanying tutorials and exercises, which relate to the real-world projects, help you understand the usage and abilities of the tools available in Autodesk Revit.

The main features of this textbook are as follows:

- **Tutorial Approach**
  The author has adopted the tutorial point-of-view approach and the learn-by-doing approach throughout the textbook. This approach helps the users learn the concepts and procedures easily.

- **Real-World Designs as Tutorials**
  The author has used real-world building designs and architectural examples as tutorials in this textbook so that the users can correlate them to the real-time designs.

- **Tips and Notes**
  Additional information related to various topics is provided to the users in the form of tips and notes.

- **Learning Objectives**
  The first page of every chapter summarizes the topics that will be covered in that chapter. This will help the users to easily refer to a topic.

Symbols Used in the Textbook

Note

The author has provided additional information to the users about the topic being discussed in the form of notes.

Tip

Special information and techniques are provided in the form of tips that help in increasing the efficiency of the users.

Formatting Conventions Used in the Textbook

Please refer to the following list for the formatting conventions used in this textbook.

• Names of tools, buttons, options, browser, palette, panels, and tabs are written in boldface.	Example: The **Beam** tool, the **Modify** button, the **Structure** panel, the **Home** tab, the **Properties** palette, **Project Browser**, and so on.
• Names of dialog boxes, drop-downs, drop-down lists, list boxes, areas, edit boxes, check boxes, and radio buttons are written in boldface.	Example: The **Options** dialog box, the **Column** drop-down in the **Structure** panel of the **Structure** tab, the **Name** edit box in the **Name** dialog box, the **Chain** check box in the **Options Bar**, and so on.
• Values entered in edit boxes are written in boldface.	Example: Enter **4'** in the **Offset** edit box.
• Names of the files saved are italicized.	Example: *c03_Industrial-Complex_tut2.rvt*
• The methods of invoking a tool/option from the ribbon, **File** menu, or the shortcut keys are given in a shaded box.	**Ribbon:** Architecture > Build > Column drop-down > Structural Column **File menu:** New **Shortcut Keys:** CTRL+N

Ribbon: Modify | (Elements / Components) > Modify > Move
Shortcut Keys: MV

Naming Conventions Used in the Textbook

Tool

If you click on an item in a panel of the ribbon and a command is invoked to create/edit an object or perform some action, then that item is termed as **tool**.

For example:
Beam tool, **Structural Column** tool, **Isolated** tool
Filled Region tool, **Trim/Extend to Corner** tool, **Rotate** tool

If you click on an item in a panel of the ribbon and a dialog box is invoked wherein you can set the properties to create/edit an object, then that item is also termed as **tool**, refer to Figure 1.

For example:
Load Family tool, **Beam** tool, **Wall** tool
Structural Wall tool, **Visibility/Graphics** tool

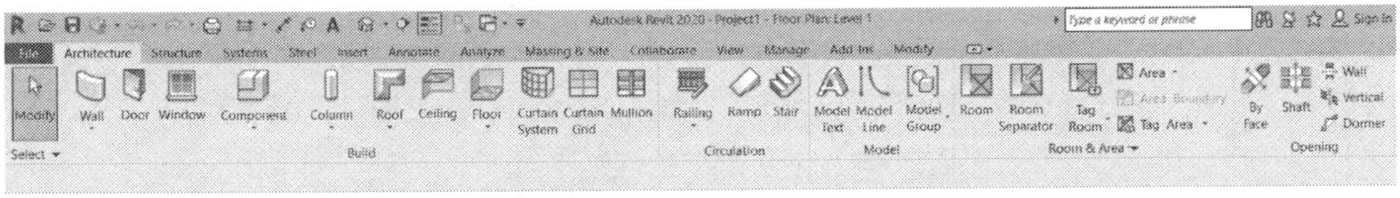

Figure 1 *Tools in the ribbon*

Button

The item in a dialog box that has a 3d shape like a button is termed as **button**. For example, **OK** button, **Cancel** button, **Apply** button, and so on. If the item in a ribbon is used to exit a tool or a mode, it is also termed as button. For example, **Modify** button, **Finish Edit Mode** button, **Cancel Edit Mode** button, and so on; refer to Figure 2.

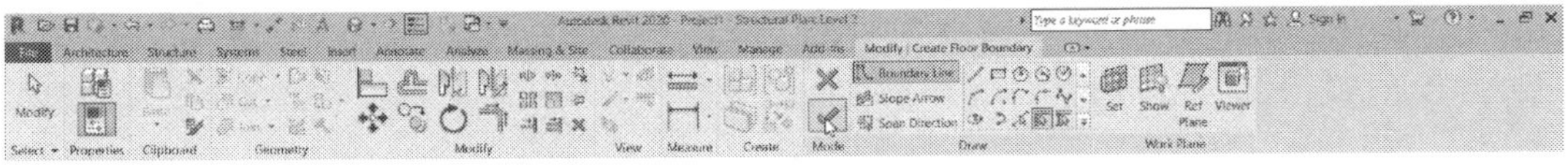

Figure 2 *Choosing the* ***Finish Edit Mode*** *button*

Dialog Box

In this textbook, different terms are used for referring to the components of a dialog box. Refer to Figure 3 for the terminology used.

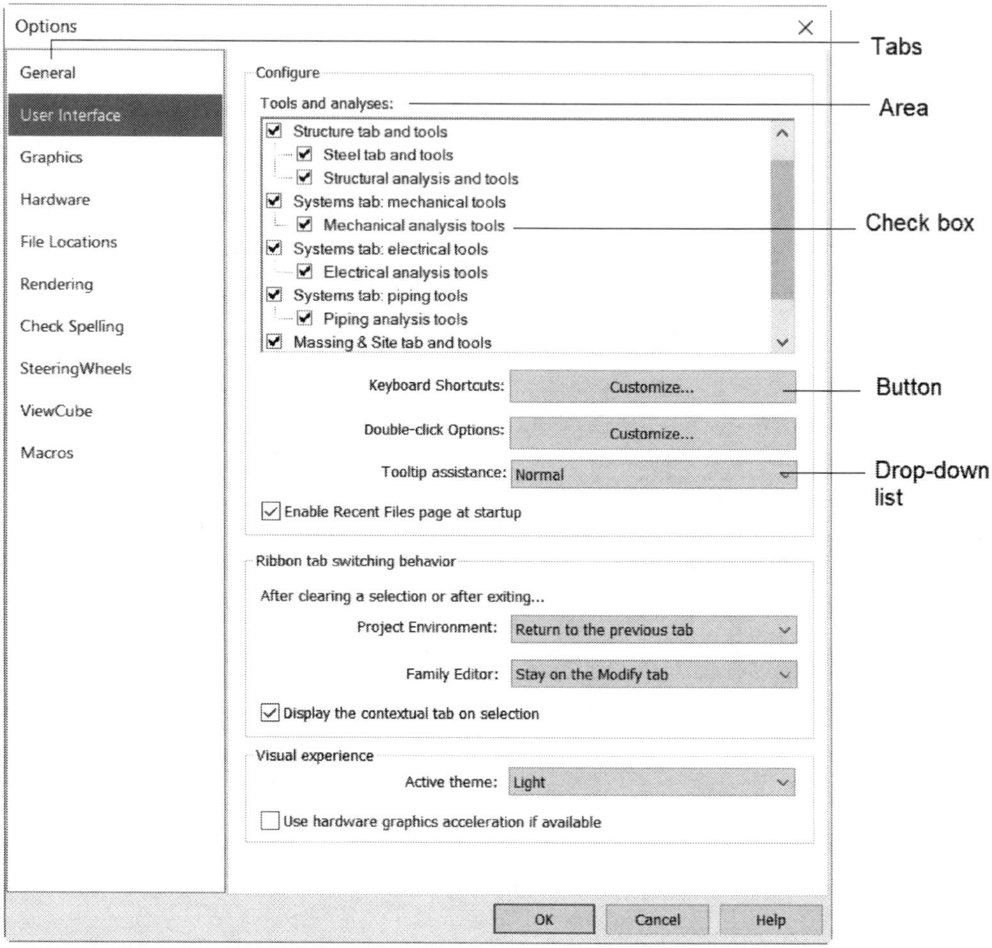

***Figure 3** The components of a dialog box*

Drop-down

A drop-down is one in which a set of common tools are grouped together for creating an object or performing some action. You can identify a drop-down with a down arrow on it. These drop-downs are given a name based on the tools grouped in them. For example, **Wall** drop-down, **Component** drop-down, **Region** drop-down, and so on; refer to Figure 4.

Drop-down List

A drop-down list is the one in which a set of options are grouped together. You can set various parameters using these options. You can identify a drop-down list with a down arrow on it. For example, **Type Selector** drop-down list, **Units** drop-down list, and so on; refer to Figure 5.

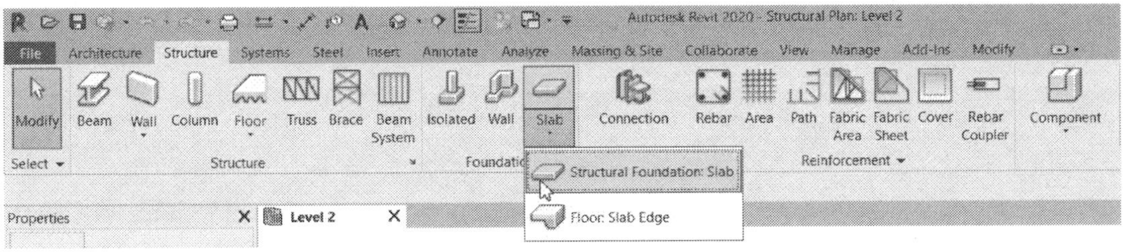

Figure 4 Choosing a tool from the drop-down

Options

Options are the items that are available in shortcut menus, drop-down lists, dialog boxes, drop-down lists, and so on. For example, choose the **Zoom In Region** option from the shortcut menu displayed on right-clicking in the drawing area; refer to Figure 6.

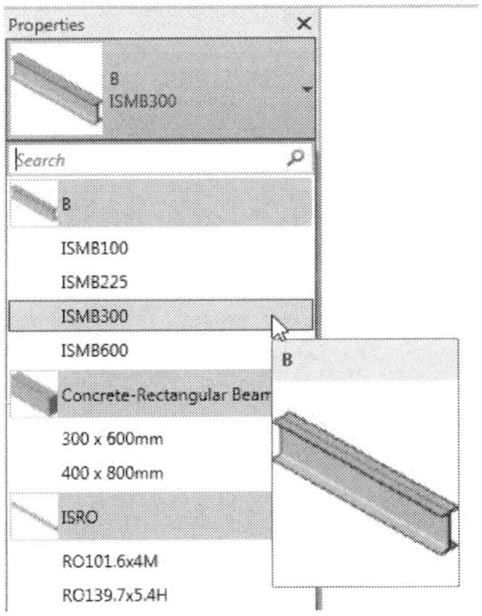

***Figure 5** Selecting an option from the **Type Selector** drop-down list*

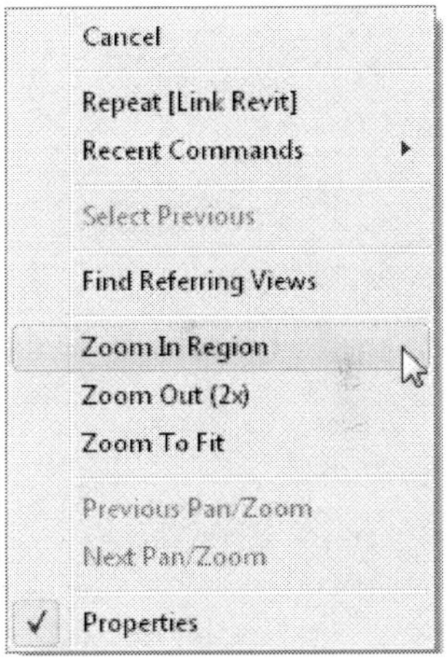

***Figure 6** Choosing an option from the shortcut menu*

This page is intentionally left blank

Chapter 1

Introduction to Autodesk Revit 2020 for Structure

Learning Objectives

After completing this chapter, you will be able to:

- *Understand the basic concepts and principles of Revit 2020*
- *Understand various terms used in Revit*
- *Describe the parametric behavior of Revit*
- *Start the Revit 2020 program*
- *Understand the interface of Revit*
- *Explain the concept of Worksharing using Revit Server*
- *Understand the linking of Analytical Model to Analysis Software*
- *Access the Revit 2020 Help*

INTRODUCTION TO Autodesk Revit for Structure

Welcome to the realm of Autodesk Revit, a powerful software for structural engineering that provides purpose-built tools for structural design, engineering, and analysis.

Autodesk Revit software provides the Engineers and Designers with tools for structural analysis and design. Revit is a BIM software that helps the users in a project to coordinate the documentation of structural design with other disciplines of engineering. Its integrated parametric modeling technology is used to create the information model of a project and to collect and coordinate information across all its representations. In Autodesk Revit, drawing sheets, 2D views, 3D views, and schedules directly represent the same building information model (BIM). In this software, the physical model is associated with an analytical model. As a result, the model created in Revit is ready to be analyzed in a compatible structural analysis software, such as Autodesk Robot Structural Analysis. Using its parametric change engine, you can modify a design at any stage of its creation. The change in the project is automatically made and represented in all its views resulting in the development of better designs along with an improved coordination. The use of Autodesk Revit provides a competitive advantage and a higher profitability to structural engineers and building industry professionals.

Autodesk Revit AS A BUILDING INFORMATION MODELER

The history of computer aided design and documentation dates back to the early 1980s when architects and engineers began using this technology for documenting their projects. Realizing its advantages, information sharing capabilities were developed especially to share data with other consultants. This led to the development of object-based CAD systems in the early 1990s. Before the development of these systems, objects such as structural walls, beams, columns, and slabs were stored as a non-graphical data with the assigned graphics. These systems arranged the information logically but were unable to optimize its usage in a building project. Realizing the advantages of the solid modeling tools, the mechanical and manufacturing industry professionals began using the information modeling CAD technology. This technology enabled them to extract data based on the relationship between model elements.

In 1997, a group of mechanical CAD technologists began working on a new software for the building industry. The Building Information Modeling (BIM) provided an alternative approach to building design, construction, and management. This approach, however, required a suitable technology to implement so as to reap its benefits. In such a situation, the use of parametric technology with the Building Information Modeling approach was envisaged as an ideal combination. They developed a software that was suitable for creating building projects. This led to the development of Autodesk Revit.

Autodesk Revit provides powerful tools for structural design and documentation, in which a digital structural model is created using the parametric elements such as structural walls, beams, columns and so on. All building elements have inherent relationship with one another, which can be tracked, managed, and maintained by the computer.

BASIC CONCEPTS AND PRINCIPLES

Autodesk Revit enables you to envisage and develop a structural model with actual 3D parametric structural elements. It provides a new approach to the structural design and the implementation process. It replicates the way structural engineers conceive the structure of a building. The 2D CAD platforms, such as AutoCAD, mostly use lines to represent the elements, as shown in Figure 1-1. However, in Autodesk Revit, you can create the structural model of a building project using 3D elements such as structural floors, columns, beams, and so on, as shown in Figure 1-2.

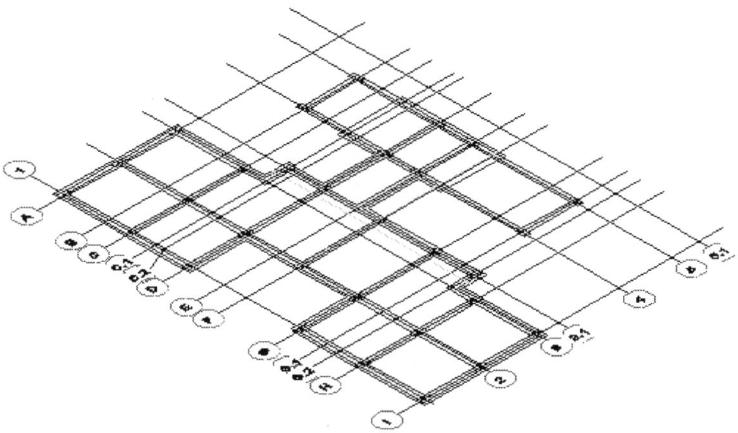

Figure 1-1 *The project created using 2D lines*

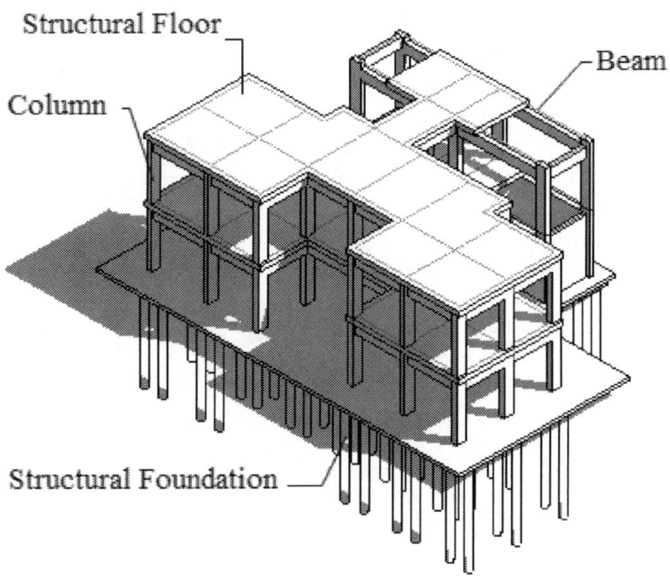

Figure 1-2 *A structural project created using parametric structural elements*

Using these 3D elements, you can visualize the structural project with respect to its scale, volume, and proportions. This enables you to study design alternatives and develop superior quality design solutions. Autodesk Revit automates routine drafting and coordination tasks and assists in reducing errors in documentation. This in turn saves time, improves the speed of documentation, and lowers the cost for users.

Understanding the Parametric Building Modeling Technology

A project in Autodesk Revit is created using the inbuilt parametric building elements. The term 'parametric' refers to the parameters that relate to various building elements. Some relationships are made by Autodesk Revit itself and others by the user. For example, doors, which have an inherent parametric relationship with the structural walls cannot be created without first creating a host wall. A door always moves with the host wall. Similarly, floors too are parametrically linked to walls. When you move structural walls, the structural floor extents are also modified automatically. Each structural element has inbuilt bidirectional associativity with many other elements in the project.

A building information model is created using different interdependent parametric building elements such as structural walls, beams, columns, structural floors, foundations, and so on. As they are bidirectionally associated elements, any change made in one element is automatically adopted by others. The integrated building information model created contains all data for a project. You can then create project presentation views such as structural plans, sections, elevations, and so on for documentation. As you modify the model while working in certain views, Autodesk Revit's parametric change engine automatically updates other views. This capability is, therefore, the underlying concept in Autodesk Revit.

Autodesk Revit's parametric change engine enables you to modify design elements at any stage of the project development. As changes are made immediately and automatically, it saves the time and effort of coordinating them in all other associated views, which, for most projects, is an inevitable part of the design process. Autodesk Revit's capability to coordinate between various aspects of the building design provides immense flexibility in the design and development process along with an error-free documentation.

Autodesk Revit also provides a variety of in-built parametric element libraries that can be selected and used to create a building model. It also provides you with the flexibility to modify the properties of these elements or to create your own parametric elements, based on the project requirement.

Terms Used in Autodesk Revit for Structure

Before using Autodesk Revit, it is important to understand the basic terms used for creating a building model. Various terms used in Autodesk Revit for Structure such as project, level, category, family, type, and instance are described next.

Autodesk Revit Project

A project in Autodesk Revit is similar to an actual structural project. In an actual project, the entire documentation such as drawings, 3D views, specifications, schedules, cost estimates, and so on are inherently linked and read together. Similarly, in Autodesk Revit, a project not only includes the digital 3D building model but also its parametrically associated documentation. Thus, all components such as the building model, its standard views, structural drawings, and schedules combine together to form a complete project. A project file contains all project information such as building elements used in a project, drawing sheets, schedules, cost estimates, 3D views, renderings, and so on. A project file also stores various settings such as environment, lighting, and so on. As data is stored in the same file, so it becomes easier for Autodesk Revit to coordinate the entire database.

Levels in a Building Model

In Autodesk Revit, a building model is divided into various levels. These levels may be understood as infinite horizontal planes that act as hosts for different elements such as roof, floor, ceiling, and so on. The defined levels in a building model can, in most cases, relate to different floor levels, or stories of the building project. Each element that you create belongs to a particular level.

Subdivisions of Elements into Categories and Subcategories

Apart from building elements, Autodesk Revit project also contains other associated elements such as annotations, imported files, links, and so on. These elements have been divided into the following categories:

Model Category : Consists of various structural elements used in creating a building model such as structural walls, structural floors, foundations, beams, braces, and columns.

Annotation Category : Consists of annotations such as dimensions, text notes, tags, symbols, and so on

Datum Category : Consists of datums such as levels, grids, reference planes, and so on

View Category : Consists of interactive project views such as structural floor plans, elevations, sections, 3D views, and renderings

In addition to these four categories, other categories such as **Imported**, **Workset**, **Filter**, and **Revit Categories** can also exist if the project has imported files, enabled worksets, or linked Autodesk Revit projects, respectively.

Families in Autodesk Revit

Another powerful concept in Autodesk Revit is family. A family is described as a set of elements of the same category that can be grouped together based on certain common parameters or characteristics. Elements of the same family may have different properties, but they all have common characteristics. For example, **Concrete-Rectangular-Column** is a concrete column family, but it contains different sizes of columns. Family files have the *.rfa* extension. You can load additional building component families from the libraries provided in Autodesk Revit package.

Families are further divided into certain types. Type or family type, as it is called, is a specific size or style of a family. For example, **Concrete-Rectangular-Column: 12 x 18** is a column type in Imperial system and **M_Concrete-Rectangular-Column: 300 x 450mm** in Metric system. All uses of the same family type in a project have the same properties. Family and family types can also be used to create new families using the **Family Editor**.

Instances are the actual usage of model elements in a building model or annotations in a drawing sheet. A family type created in a new location is identified as an instance of the family type. All instances of the same family type have the same properties. Therefore, when you modify the properties of a family type, the properties of all its instances also get modified. The family categorization of Revit elements is given below:

Model Category : Column
Family : **Concrete-Rectangular-Column** in Imperial
M_Concrete-Rectangular-Column in Metric
Family type : 12 x 18 (300 x 450mm)
Instance : Particular usage of a family type

The hierarchy of building elements in Autodesk Revit plays an important role in providing the flexibility and ease of managing a change in a building model. Figure 1-3 shows the hierarchy of categories and families in a typical Autodesk Revit project.

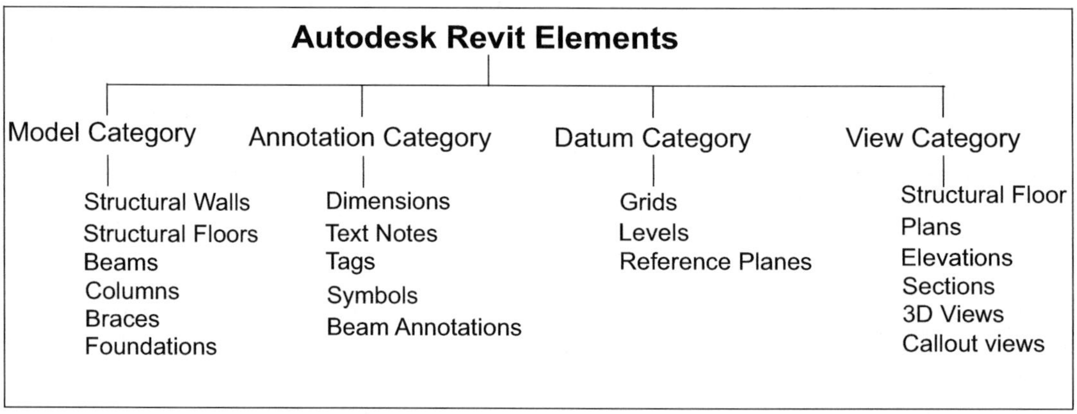

***Figure 1-3** Hierarchy of Autodesk Revit categories and families*

Creating a Structural Model Using Parametric Building Elements

Another classification of categories of elements followed in Autodesk Revit is based on their usage. Autodesk Revit uses five classes of elements: Host, component, annotation, view, and datum. Hosts are the element categories that form the basic structure of a building model and include model elements such as structural walls and floors. Components are the elements that are added to the host elements or act as stand-alone elements such as doors, windows, and foundations. Annotations are the 2D, view-specific elements that add content to the project documentation such as dimensions, tags, text notes, and so on. Views represent various orientations of a building model such as plans, elevations, sections, 3D views, and so on. Datum refers to the reference elements that assist you in creating a building model. The reference elements which include grids, levels, reference planes, and so on.

There is no specific methodology available for creating a building model in Autodesk Revit. It provides you with the flexibility of generating the building geometry based on the project requirement, design complexity, and other factors. However, the following steps describe a general procedure that may be followed for creating an architectural building model using the built-in parametric elements provided in Autodesk Revit.

The first step is to define the levels of the structural model based on the story height of the building and then create grids for inserting columns and foundation at the lowest level. Next, add columns, foundation slab, structural wall, and foundations in that level. You can also link the control height of the structural walls and columns to the levels. Next, create framing members and floors using the defined levels. You can add loads to the model and define load conditions and various analytical settings for the model.

After creating the structural analytical model, you will transfer it to Autodesk Robot Structural Analysis software and analyze the structural entities based on the loads applied to it. After performing the analysis and retrieving the design detail of the model, you will import the structural model from the analysis software into Autodesk Revit. Next, you will create drawing sheets with the desired views for its presentation. You can also add reinforcements to the concrete elements in the structural model. Autodesk Revit also provides tools to create rendered 3D views and walkthroughs. Figure 1-4 shows an example of a building elevation with various structural elements.

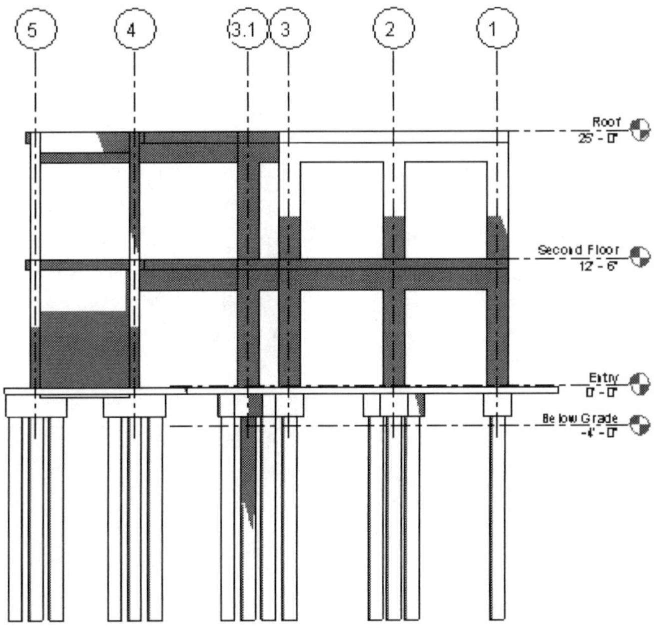

Figure 1-4 *Building section showing building elements and levels*

Visibility/Graphics Overrides, Scale, and Detail Level

Autodesk Revit enables you to control the display and graphic representation of a single element or the element category of various elements in project views by using the visibility and graphics overrides tools. You can select a model category and modify its linetype and detail level. This can also be done for various annotation category elements and imported files. These settings can be done for each project view based on its desired representation. You can also hide an element or an element category in a view using the **Hide in view** and **Isolate** tools. You can override the graphic representation of an element or an element category in any view using the **Visibility/ Graphics** tool.

The scale is an important concept in a Revit project. You can set the scale from the available list of standard scales such as 1/16"=1'0", 1/4"=1'0", 1"=1'0", 1/2"=1'0" (for Imperial) and 1:1, 1:2, 1:3, 1:4, 1:5, 1:10 (for Metric). As you set a scale, Autodesk Revit automatically sets the detail level that is appropriate for it. There are three detail levels provided in an Autodesk Revit project: **Coarse**, **Medium**, and **Fine**. You can also set the detail level manually for each project view. Each detail level has an associated linetype and the detail lines associated with it. The details of annotations such as dimensions, tags, and so on, are also defined by the selected scale.

Extracting Project Information

A single integrated building information is used to create and represent a building project. You can extract project information from a building model and create area schemes, schedule, and cost estimates, and then add them to the project presentation.

Autodesk Revit also enables you to export the extracted database to the industry standard Open Database Connectivity (ODBC) compliant relational database tables. The use of the building information model to extract database information eliminates the error-prone method of measuring building spaces individually.

Creating a Structural Drawing Set

After creating the building model, you can easily arrange the project views by plotting them on the drawing sheets. Drawing sheets can also be organized in a project file based on the established CAD standards followed by the firm. In this manner, the project documentation can easily be transformed from the conceptual design stage to the design development stage and finally to the construction document stage. The project view on a drawing sheet is only a graphical representation of the building information model. Therefore, any modification made in it is immediately updated in all the associated project views, thereby keeping the drawing sets updated.

Creating an Unusual Building Geometry

Autodesk Revit also helps you conceptualize a building project in terms of its volume, shape, and proportions before working with actual building elements. This is possible by using the Massing tool, which enables you to create quick 3D models of buildings and conduct volumetric and proportion study on overall masses. It also enables you to visualize and create an unusual building geometry. The same massing model can then be converted into a building model with individual parametric building elements. It provides continuity in the generation of building model right from sketch design to its development.

Flexibility of Creating Special Elements

Autodesk Revit provides a large number of in-built family types of various model elements and annotations. Each parametric element has the associated properties that can be modified based on the project requirement.

Autodesk Revit also enables you to create the elements that are designed specifically for a particular location. The in-built family editor enables you to create new elements using family templates. This provides you with the flexibility of using in-built elements for creating your own elements. For example, using the furniture template, you can create a reception desk that is suitable for a particular location in the design.

Creating Structural Layouts

Autodesk Revit's structural tools enable you to add structural elements to a building model. An extensive in-built library of structural elements has been provided in Autodesk Revit. You can add structural columns, beams, walls, braces, and so on to the project. Thus, structural consultants can also incorporate their elements in the basic architectural building model and check for inconsistency, if any.

Working on Large Projects

In Autodesk Revit, you can work on large projects by linking different building projects together. For a large project that consists of a number of buildings, you can create individual buildings as separate projects and then link all of them into a single base file. The database recognizes the linked projects and includes them in the project representation of the base file.

For example, while working on a large educational institution campus, you can create separate project files for academic building, administration area, gymnasium, cafeteria, computer centre, and so on, and then link them into the base site plan file. In this manner, large projects can be subdivided and worked upon simultaneously.

Working in Large Teams and Coordinating with Consultants

Worksets, in Autodesk Revit, enable the division of the building model into small editable set of elements. The worksets can be assigned to different teams working on the same project and then their work can easily be coordinated in the central file location. The effort required to coordinate, collaborate, and communicate the changes between various worksets is taken care of by computer. Various consultants working on a project can be assigned a workset with a set of editable elements. They can then incorporate their services and modify the associated elements.

For example, a high rise commercial building project can be divided into different worksets with independent teams working on exterior skin, interior walls, building core, toilet details, finishes, and so on. The structural consultants can be assigned the exterior skin and the core workset, in which they can incorporate structural elements. Similarly, the rest of the teams can work independently on different worksets.

STARTING Autodesk Revit 2020

You can start Autodesk Revit by double-clicking on its shortcut icon on the desktop. Alternatively, you can start Autodesk Revit 2020 from the taskbar. To do so, choose the **Start** button; a menu is displayed, as shown in Figure 1-5. Choose **Autodesk** and then **Revit 2020**; the interface will be displayed, as shown in Figure 1-6. (For Windows 10)

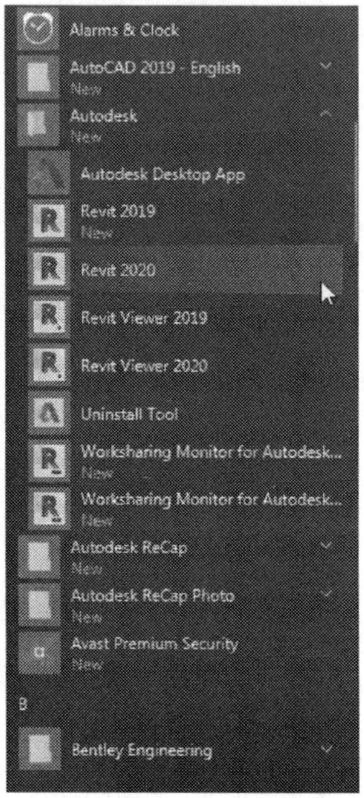

***Figure 1-5** Starting Autodesk Revit 2020 using the taskbar*

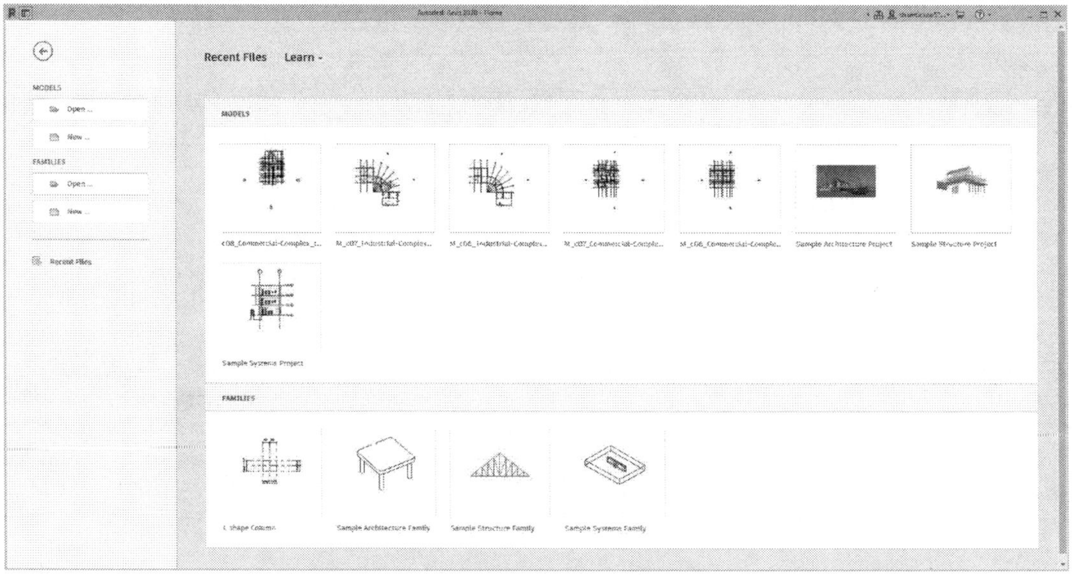

***Figure 1-6** Autodesk Revit 2020 interface*

Note
The path for starting Autodesk Revit depends on the operating system being used.

The screen interface has three sections: **Projects**, **Families**, and **Resources**. The options in the **Projects** section are used to open a new or an existing project. In the **Projects** area, you can add some of the frequently used templates such as the the **Construction** template, just below the **New** option. To do so, choose the **File** menu and then choose the **Options** button from the flyout displayed. On doing so, the **Options** dialog box will be displayed. In this dialog box, choose the **File Locations** tab and then choose the **Add Value** button; the **Browse for Template File** dialog box will be displayed. In this dialog box, select the required template from the appropriate location and choose the **Open** button; the template will be added under the **Name** column and will also be available under the **Projects** area. Similarly, options in the **Families** section are used to open a new or an existing family. You can also invoke the Conceptual Mass environment from this section to create a conceptual mass model.

In the **Resources** section, you can choose the **What's New?** option to get information about the new tools and features in Revit 2020. In addition, you can choose the **Help** option from the **Resources** section to get help on various tools. When you choose this option, you will be directed to ***https://help.autodesk.com/view/RVT/2020/ENU/***. Also, the **Autodesk Revit 2020** page with the **Welcome to Revit 2020 Learning** area will be displayed. To access information related to additions and enhancements in Revit 2020 release, you can expand the **What's New** node from the left pane and then click on the **What's New** link. On doing so, the **Whats's New** page will be displayed in the right pane with various links. You can get information about various enhancements in Revit 2020 by accessing these links.

The **What's New? Videos** option in the **Resources** section is an enhancement in Revit 2020. When you click on this option, you are directed to the **Autodesk Revit 2020** page that has a list of videos of newly added features in Revit. You can click on a link to view the corresponding video.

You can choose the **Essential Skills Videos** and **Additional Videos** options to view the videos related to basic and advance concepts in Revit 2020. These videos and their associated information help you to learn about the complete software. Moreover, you can choose the **Autodesk App Store** option to access various add-ons that can be used to enhance the productivity of Revit. On choosing this option, the **AUTODESK APP STORE** page will be displayed. In this page, various links are available as add-ons which can be used in Revit applications. In the **Revit Community** option of the **Resources** section, you can access information related to various communities and their contribution in the form of articles, tutorials, and videos.

In the **Projects** section, choose the **Open** option; the **Open** dialog box will be displayed. Browse to the desired location in the dialog box and select the file. Now, choose the **Open** button to open the file.

To open a new project file, choose the **New** option from the **Projects** section. Alternatively, choose **New > Project** from the **File** menu; the **New Project** dialog box will be displayed. In this dialog box, make sure that the **Project** radio button is selected, and then choose the **OK** button; a new project file will open and the interface screen will be activated.

To create a new project template file, choose the **New** option from the **Projects** section; The **New Project** dialog box will be displayed. In this dialog box, select the base template from the **Template file** drop-down list and make sure that the **Project template** radio button is selected. Choose the **OK** button; a new project template file will open with the interface screen.

USER INTERFACE

In Autodesk Revit, the user interface consists of the ribbon, Drawing Area, **Properties** palette, Status Bar, and the **View Control Bar**, as shown in Figure 1-7. In Autodesk Revit, ribbon is an interface from where you can invoke tools. The ribbon, which contains task-based tabs and panels, streamlines the structural workflow and optimizes the project delivery time. In Autodesk Revit, when you select an element in the drawing area, the ribbon displays a contextual tab that comprises of tools corresponding to the selected element. The interface of Autodesk Revit is similar to the interfaces of many other Microsoft Windows-based programs. The main parts in the Revit interface are discussed next.

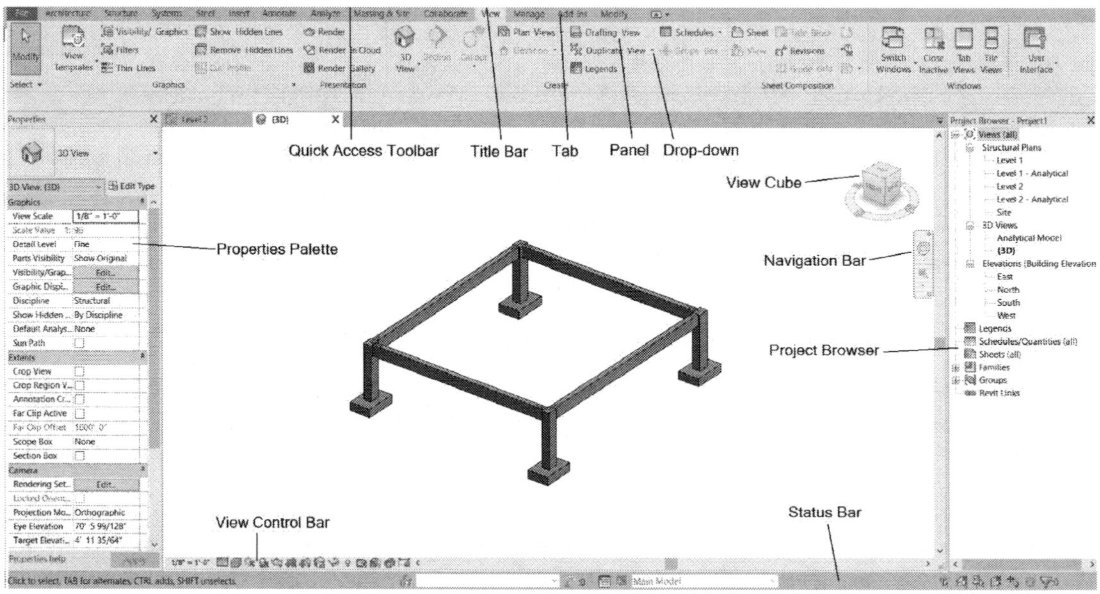

Figure 1-7 *The Autodesk Revit 2020 user interface*

Title Bar

The Title Bar is displayed on the top portion in the user interface, displays the program's logo, name of the current project, and the view opened in the viewing area.

Ribbon

The ribbon, as shown in Figure 1-8, is an interface that is used to invoke tools. When you open a file, the ribbon is displayed at the top in the screen. It comprises task-based tabs and panels, refer to Figure 1-8, which provide all the tools necessary for creating a project. The tabs and panels in the ribbon can be customized according to the need of the user. This can be done by moving the panels and changing the view states of the ribbon (the method of changing the ribbon view state is discussed later in this chapter). The ribbon contains buttons, drop-downs, panels, tabs, and tools. These buttons and the tools can be selected from the corresponding panels.

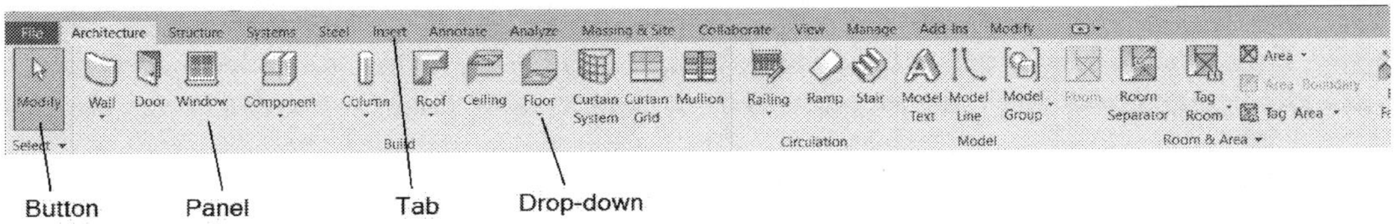

Figure 1-8 *Different components of a ribbon*

Tip
Tooltips appear when you place the cursor over any of the tool icons in the ribbon. The name of the tool appears in a box helping you in identifying the tool icon.

In the ribbon, you can move a panel and place it anywhere on the screen. To do so, press and hold the left mouse button on the panel label in the ribbon, and then drag the panel to a desired place on the screen. Next, use the tools of the moved panel and place the panel back in the ribbon. To do so, place the cursor on the moved panel and choose the **Return Panels to Ribbon** button from the upper right corner of this panel, as shown in Figure 1-9; the panel will return to the ribbon.

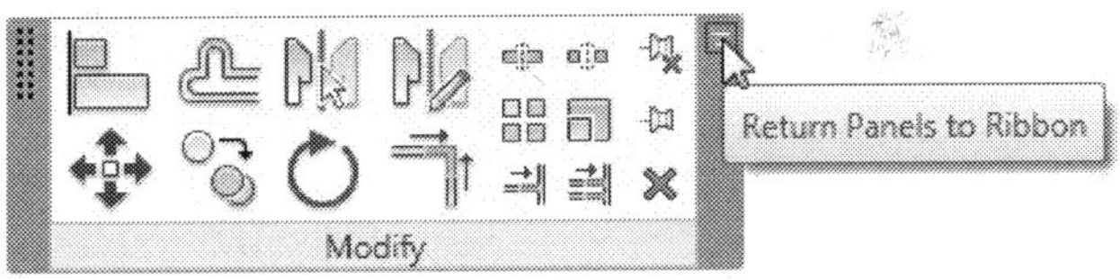

Figure 1-9 *Choosing the* ***Return Panels to Ribbon*** *button*

Changing the View States of the Ribbon

The ribbon can be displayed in three view states by selecting any of the following four options: **Minimize to Tabs**, **Minimize to Panel Titles**, **Minimize to Panel Buttons**, and **Cycle through All**. To use these options, place the cursor over the second arrow on the right of the **Modify** tab in the ribbon, refer to Figure 1-10; the arrow will be highlighted. Next, click on the down arrow; a flyout will be displayed, as shown in Figure 1-10. From this flyout, you can choose the **Minimize to Tabs** option to display the only tabs in the ribbon. If you choose the **Minimize to Panel Titles** option, the ribbon will display the titles of the panels along with the tabs. You can choose the **Minimize to Panel Buttons** option to display panels as buttons in the ribbon along with tabs.

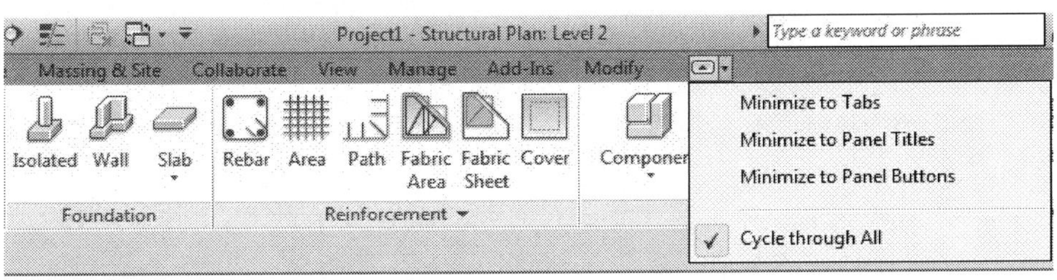

Figure 1-10 *Various options in the flyout for changing the view state of the ribbon*

Note

*If the ribbon is changed to a different view state, then on placing the cursor over the first arrow on the right of the **Modify** tab, the **Show Full Ribbon** tooltip will be displayed. Click on the arrow; the full ribbon will be displayed.*

The table given next describes the tabs in the ribbon and their functions.

Tab	**Description**
Architecture	Contains tools for creating an architectural model
Structure	Contains tools for creating a structural model
Systems	Contains various tools for HVAC, Mechanical, Plumbing, and Electrical purposes
Steel	Contains tools for creating connections, fabricating elements, and modifying the edges
Insert	Contains tools for inserting or managing secondary files such as raster image files and CAD files
Annotate	Contains tools for documenting a building model such as adding texts and dimensions
Analyze	Contains tools for analyzing the structural model
Massing & Site	Contains tools for creating massing and site elements
Collaborate	Contains tools for collaborating the project with other team members (internal and external)
View	Contains tools for managing and modifying the current views, switching views, and so on
Manage	Contains tools for specifying the project and system parameters and settings
Add-Ins	Contains add ins tools available to subscription customers to use it with the Revit software
Modify	Contains tools for editing elements in the model

Contextual Tabs in the Ribbon

These tabs are displayed when you choose certain tools or select elements. They contain a set of tools or buttons that relate only to a particular tool or element.

For example, when you invoke the **Beam** tool, the **Modify | Place Beam** contextual tab is displayed. This tab has the following panels: **Select**, **Properties**, **Clipboard**, **Geometry**, **Modify**, **View**, **Measure**, **Create**, **Mode**, **Draw**, **Multiple**, and **Tag**. The **Select** panel contains the **Modify** tool. The **Properties** panel contains the **Properties** button and the **Type Properties** tool. The **Mode** panel has some necessary tools that are used to load model families or to create the model of a window in a drawing. The other panels, apart from those discussed above, contain the tools that are contextual and are used to edit elements when they are placed in a drawing or selected from a drawing for modification.

Application Frame

The application frame helps you manage projects in Autodesk Revit. It consists of the **File** menu, Quick Access Toolbar, InfoCenter, and Status Bar. These options are discussed next.

File Menu

The **File** menu contains tools that provide access to many common file actions such as **Open**, **Close**, and **Save**. To display this menu, choose **File** menu in the ribbon, refer to Figure 1-11. Alternatively, press ALT+F to display tools in the **File** menu.

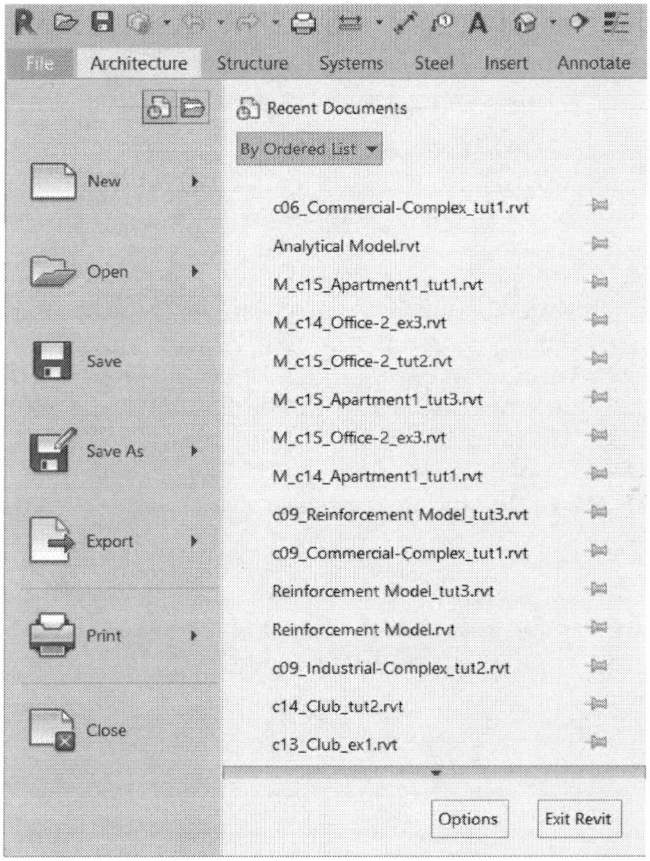

***Figure 1-11** The **File** menu*

Quick Access Toolbar

The **Quick Access Toolbar**, shown in Figure 1-12, contains the options to undo and redo changes, open and save a file, create a new file, and so on.

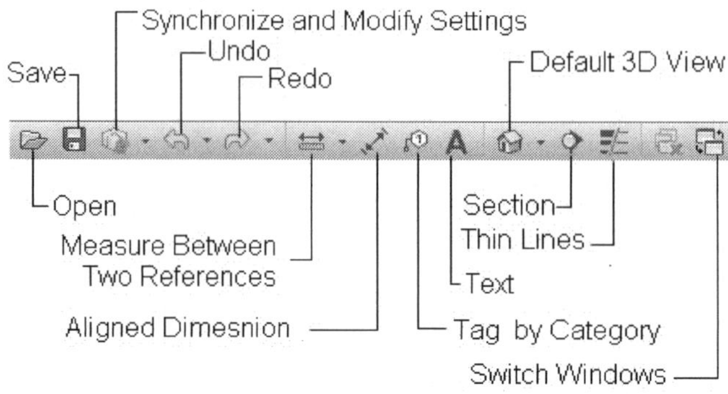

*Figure 1-12 The **Quick Access Toolbar***

By default, the **Quick Access Toolbar** contains the options such as **Open**, **Save**, **Redo**, **Undo**, and so on. You can customize the display of the **Quick Access Toolbar** by adding more tools and removing the unwanted tools. To add a tool or a button from the panel of the ribbon to the **Quick Access Toolbar**, place the cursor over the button; the button will be highlighted. Next, right-click; a flyout will be displayed. Choose **Add to Quick Access Toolbar** from the flyout displayed; the highlighted button will be added to the **Quick Access Toolbar**. The **Quick Access Toolbar** can be customized to reorder the tools displayed in it. To do so, choose the down arrow next to the **Switch Windows** drop-down, refer to Figure 1-12; a flyout will be displayed. Choose the **Customize Quick Access Toolbar** option located at the bottom of the flyout; the **Customize Quick Access Toolbar** dialog box will be displayed. Use various options in this dialog box and choose the **OK** button; the **Customize Quick Access Toolbar** dialog box will close and the tools in the **Quick Access Toolbar** will be reordered.

InfoCenter

You can use **InfoCenter** to search for information related to Revit (Help) to display the **Subscription Center** panel for subscription services and product updates, and to display the **Favorites** panel to access saved topics. Figure 1-13 displays various tools in **InfoCenter**.

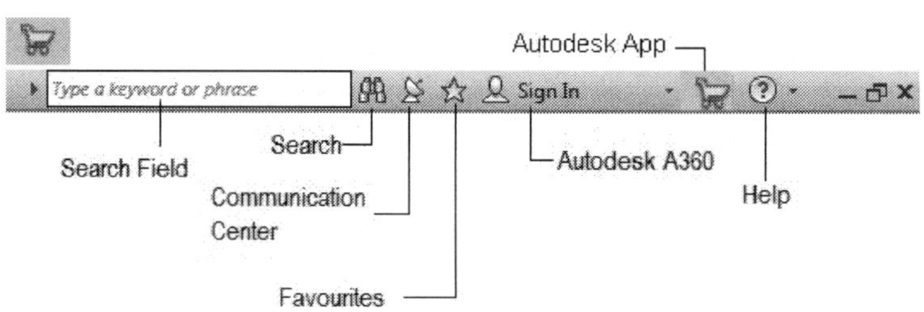

*Figure 1-13 The **InfoCenter***

Status Bar

The Status Bar is located at the bottom of the interface screen. When the cursor is placed over an element or a component, the Status Bar displays the name of the family and type of the corresponding element or components. It also displays prompts and messages to help you use the selected tools.

View Control Bar

The **View Control Bar** is located at the lower left corner of the drawing window, as shown in Figure 1-14. It can be used to access various view-related tools. The **Scale** button shows the scale of the current view. You can choose this button to display a flyout that contains standard drawing scales. From this flyout, you can then select the scale for the current view. The **Detail Level** button is used to set the detail level of a view. You can select the required detail level as **Coarse**, **Medium**, and **Fine**. Similarly, the **Visual Style** button enables you to set the display style. The options for setting the display style are: **Wireframe**, **Hidden Line**, **Shaded**, **Consistent Colors**, **Realistic**, and **Raytrace**.

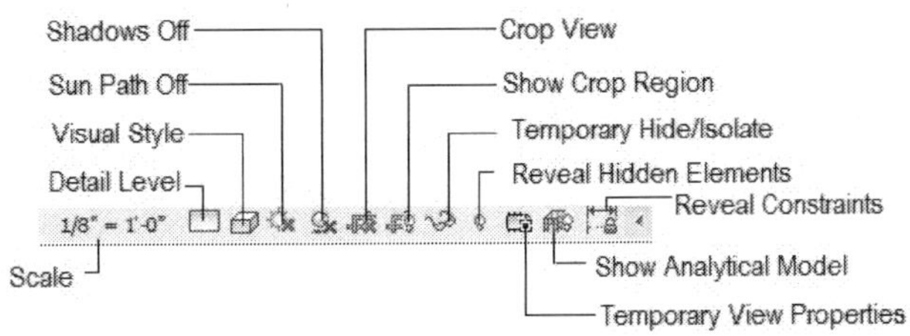

*Figure 1-14 The **View Control Bar***

Options Bar

The **Options Bar** provides information about the common parameters of a component type. It also displays options for creating or editing them. The options displayed in the **Options Bar** depend on the type of component being created and selected for editing. Figure 1-15 displays the options in the **Options Bar** to create a structural column.

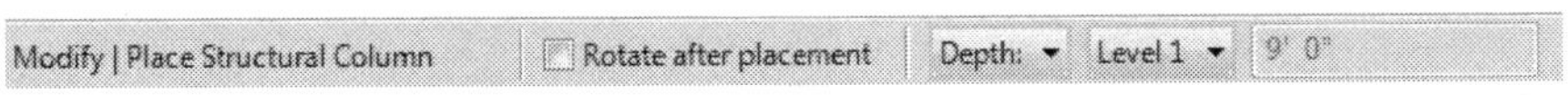

*Figure 1-15 The **Options Bar** with different options to create a structural column*

Type Selector

The **Type Selector** drop-down list is located in the **Properties** palette for the currently invoked tool. On invoking the **Beam** tool, the properties of the beam will be displayed in the **Properties** palette. In this palette, you can use the **Type Selector** drop-down list to select the required type of the beam. The content in the **Type Selector** drop-down list keep changing, depending upon the current function of the tool or the elements selected. The **Type Selector** drop-down list can also be used to specify the type of the element or component while placing an element or a component in a drawing. You can also use this drop-down list to change the type of a selected element.

Drawing Area

The Drawing Area is the actual modeling area where you can create and view the building model. It covers the major portion of the interface screen. You can draw building components in this area. The position of the pointing device is represented by the cursor. The Drawing Area also has the standard Microsoft Windows functions and buttons such as close, minimize, maximize, scroll bar, and so on. These buttons have the same function as that of the other Microsoft Windows-based programs.

PROJECT BROWSER

The **Project Browser** is located below the ribbon. It displays project views, schedules, sheets, families, and groups in a logical tree-like structure, as shown in Figure 1-16 and helps you open and manage them. To open a view, double-click on the name of the view, or drag and drop the view in the Drawing Area. You can close the **Project Browser** or dock it anywhere in the Drawing Area.

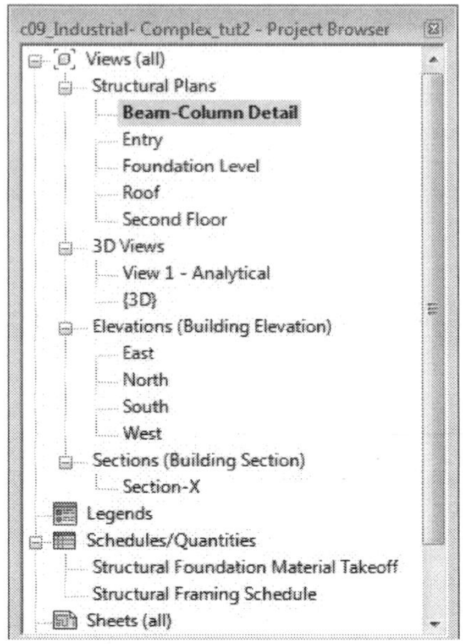

*Figure 1-16 The **Project Browser***

Note

*If the **Project Browser** is not displayed on the screen,choose the **View** tab from the ribbon and then select the **Project Browser** check box from **View > Windows > User Interface** drop-down.*

The **Project Browser** can be organized to group the views and sheets based on the project requirement. For example, while working on a large project with a number of sheets, you can organize the **Project Browser** to view and access specific sheets.

Note

*In the **Project Browser**, you can expand or collapse the view listing by selecting the '+' or '-' sign, respectively. The current view in the drawing window is highlighted in bold letters. The default project file has a set of preloaded views.*

Keyboard Accelerators

In Autodesk Revit, accelerator keys have been assigned to some of the frequently used tools. These keys are shortcuts that you can type through the keyboard to invoke the corresponding tool. Accelerator keys corresponding to a tool appear as a tooltip when you move the cursor over the tool.

Properties Palette

The **Properties** palette, as shown in Figure 1-17, is a modeless interface, which displays the type and element properties of various elements and views in a drawing. The **Properties** palette is dockable and resizable, and it supports multiple monitor configurations. The **Properties** palette is displayed in the Revit interface by default and it shows the instance properties of an active view. When you select an element from a drawing, the **Properties** palette displays its instance properties. You can also access the type properties of the selected element from the **Properties** palette. To do so, choose the **Edit Type** button from the palette; the **Type Properties** dialog box will be displayed. In this dialog box, you can change the type properties of the selected element.

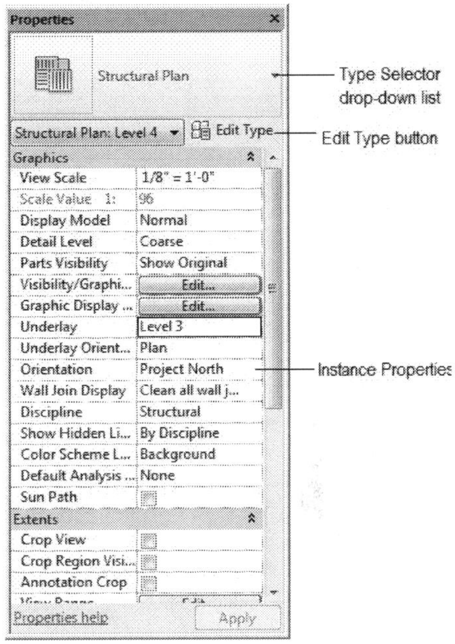

*Figure 1-17 The **Properties** palette*

In the **Properties** palette, you can assign a type to a selected element in a drawing from the **Type Selector** drop-down list. In Revit, you can toggle the display of the **Properties** palette in its interface. Choose the **Properties** button in the **Properties** panel of the **Modify** tab to hide it. Similarly, you can choose the **Properties** button to display the palette if it is not visible in the interface.

> **Tip**
> *As you become accustomed to use Autodesk Revit, you will find these **Keyboard Accelerators** quite useful because they save the effort of browsing through the menus.*

DIALOG BOXES

Some Autodesk Revit tools, when invoked, display a dialog box. A dialog box is an interface for accessing, specifying, and modifying the parameters related to that tool. For example, when you choose **Save As > Project** from the **Application Menu**, the **Save As** dialog box is displayed, as shown in Figure 1-18. A dialog box consists of various parts such as dialog label, radio buttons, text or edit boxes, check boxes, slider bars, image box, buttons, and tools, which are similar to other windows-based programs. Some dialog boxes contain the [**...**] button, which displays another related dialog box. There are certain buttons such as **OK**, **Cancel**, and **Help**, which appear at the bottom of most of the dialog boxes. The names of the buttons imply their respective functions. The button with a dark border is the default button.

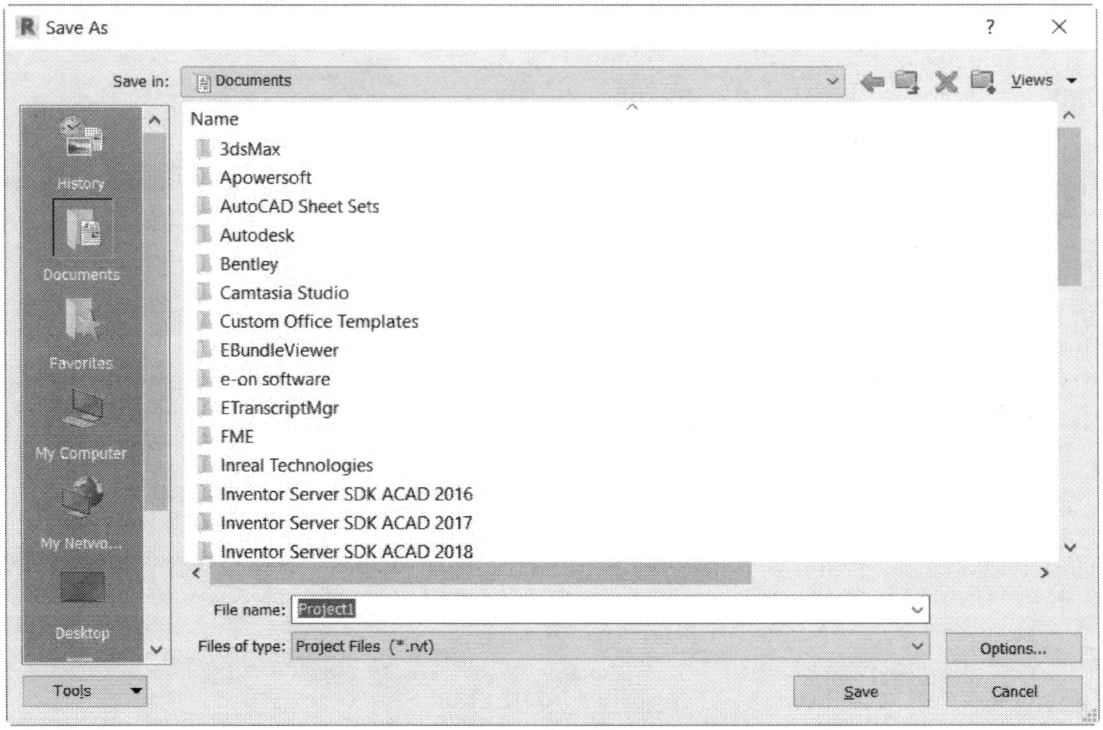

***Figure 1-18** The **Save As** dialog box*

MULTIPLE DOCUMENT ENVIRONMENT

The multiple document environment feature allows you to open more than one project at a time in a single Autodesk Revit session. This is very useful when you want to work on different projects simultaneously and make changes with reference to each other.

Sometimes, you may need to incorporate certain features from one project into the other. With the help of multiple document environment feature, you can open multiple projects and then use the **Cut**, **Copy** and **Paste** tools from the **Clipboard** panel of the **Modify (type of element)** tab to transfer the required components from one project to another. These editing tools can also be invoked by using the CTRL+C and CTRL+V keyboard shortcuts.

To access the opened projects, click on the **Switch Windows** drop-down list in the **Windows** panel of the **View** tab; the options for the names of different opened project files will be displayed, as shown in Figure 1-19. Like other Microsoft Windows-based programs, you can select and view the opened projects using the **Tile Views** tools from the **Windows** panel of the **View** tab. The tile views of a project is shown in Figure 1-20.

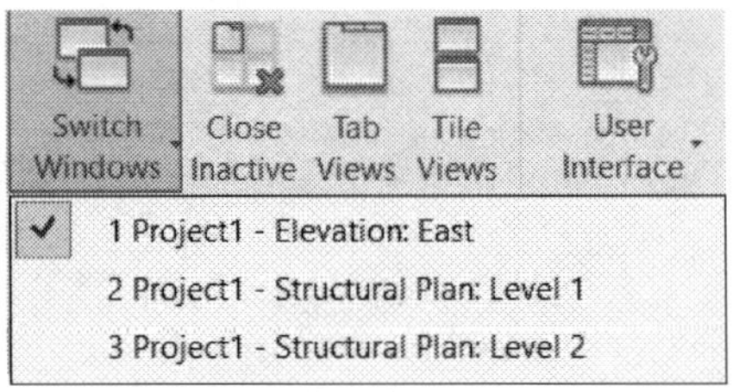

***Figure 1-19** Selecting an option from the **Switch Windows** drop-down list*

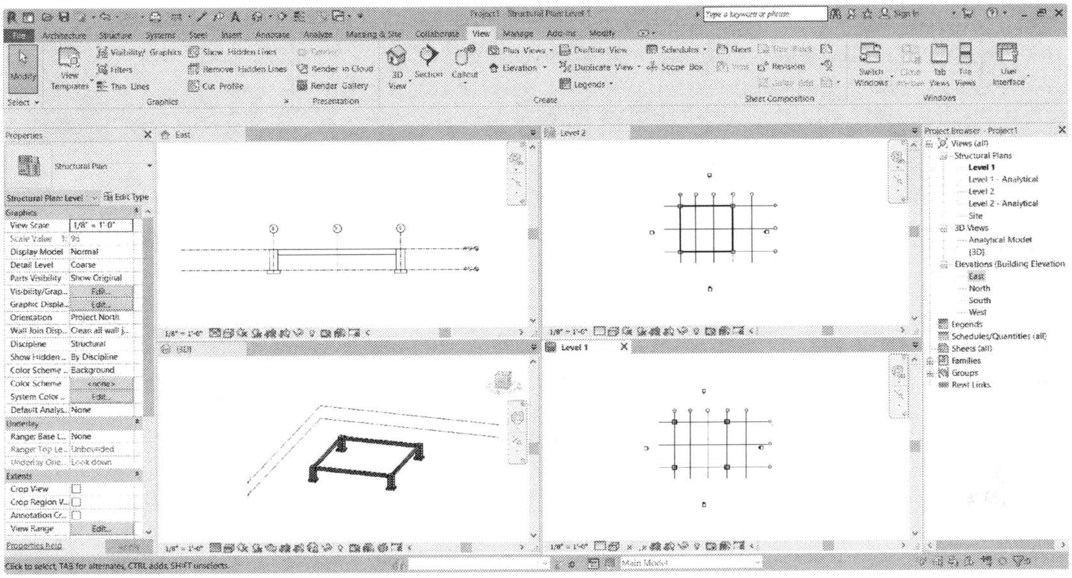

***Figure 1-20** The **Tile Views** tool for project*

INTEROPERABILITY OF Autodesk Revit

The models or geometries created in Autodesk Revit can be easily exported to AutoCAD and AutoCAD Architecture in the DWG file format. This enables structural engineers to collaborate with Architects.

One of the important aspects of the job of a structural engineer is to collaborate and share information with the rest of the design team including the architect. To facilitate this requirement, Revit follows a wide range of industry standards and supports various CAD file formats such as DWF, DGN, DWG, DGN, IFC, SKP, and SAT. For image files, it supports JPG, TIFF, BMP, PNG, AVI, PAN, IVR, and TGA file formats. Besides these, the formats that are supported by Revit include ODBC, HTML, TXT, XML, XLS, and MDB. Autodesk Revit is compatible with any CAD system that supports the DWG, DXF, or DGN file format. Revit can import the models and geometries as ACIS solids. This enables engineers to import models from AutoCAD Architecture and AutoCAD MEP (Mechanical, Electrical, and Plumbing) software and to link and import 3D information to Revit. This feature makes Autodesk Revit 2020 an efficient, user-friendly, and compatible software.

BUILDING INFORMATION MODELING AND Autodesk Revit

Building Information Modeling (BIM) is defined as a design technology that involves the creation and use of coordinated, internally consistent, and computable information about a building project in design and construction.

Using BIM, you can demonstrate the entire life cycle of a building project starting from the process of construction, facility operation, and information about quantities and shared properties of elements. BIM enables the circulation of virtual information model from the design team to contractors and then to the owner, thereby adding changes and their knowledge to update the model at each stage of transfer. The ability to keep information up-to-date and make it available in an integrated digital environment enables the architects, owners, builders, and engineers to have clear vision of the project before the commencement of actual construction. It also enables them to make better and faster decisions to improve the quality and profitability of projects. Autodesk Revit is a specially designed platform based on BIM.

In Revit, the analytical and physical representations of a structural model are created simultaneously. These representations are just different views of a computable building model that contains necessary information for a third-party analysis application. Revit provides a common modeling interface for third-party analysis applications. You can use Revit API to move data directly from the Revit building information model to the analysis software. You can further bring back the analysis while keeping the analysis, design, and documentation synchronized.

Revit's parametric model represents a building as an integrated database of coordinated information. In Revit, change anywhere is change everywhere. A change made in your project at any stage is reflected in the entire project, and also, due to the parametric behavior of elements, the project is updated automatically. Also, the integration of Revit with the available in-built commercial tools such as solar studies, material takeoffs, and so on greatly simplifies the project design and reduces the time consumed by these analyses, thereby enabling faster decision making.

WORKSHARING USING REVIT SERVER

Worksharing is a method of distributing work among people involved in a project, and accomplishing it within the stipulated period of time. In worksharing, each person involved in the project is assigned a task that has to be accomplished through proper planning and by coordinating with the other members of the team.

In a large scale building project, worksharing helps in finishing a project in time and meeting the quality requirements that are set during the process. Generally, in a large scale building project, worksharing is based on the specialization of work. The professionals such as structural engineers, architects, interior architects, and MEP engineers are involved in their respective fields to accomplish the project. So, the distribution of work at the primary stage is made on the basis of the area of specialization. Each professional has his own set of work to perform for the accomplishment of the project. Therefore, worksharing is an important process that is required to be implemented efficiently to complete the project in time.

In Autodesk Revit, you can apply server-based worksharing with the help of Revit Server, which is a server application. Revit Server uses a central server and multiple local servers for collaborating across a Wide Area Network (WAN). The central server hosts the central model of a workshared project and remains accessible to all team members over the Wide Area Network. Similarly, the local server is accessible to all team members in a Local Area Network (LAN). The local server hosts a local updated copy of the central model. In the Worksharing environment, the team members are not aware of the local server as it is transparent in their daily operations. Refer to Figure 1-21 for the network model of Revit Server.

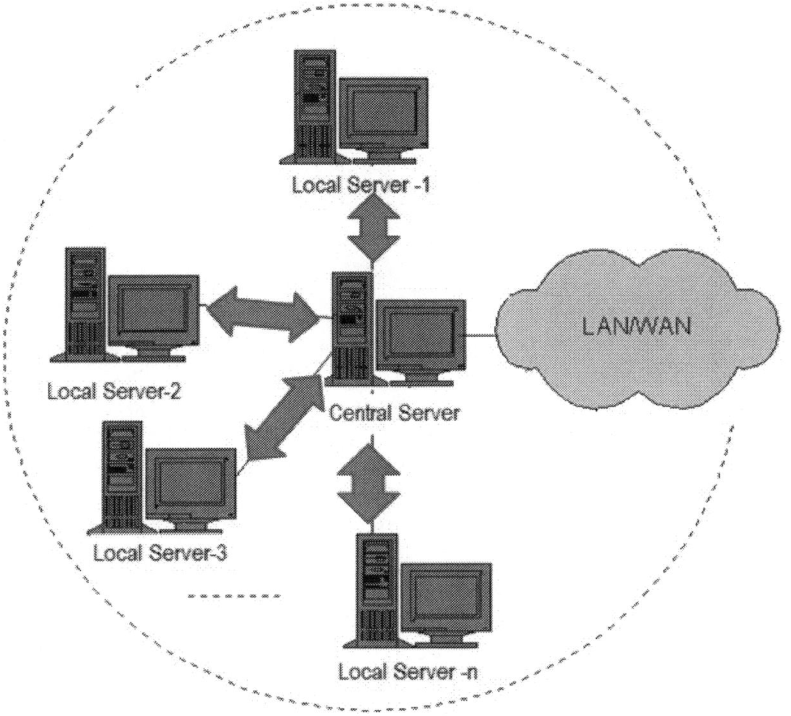

Figure 1-21 *The Network Model of Revit Server*

In Worksharing environment, a team member starts working on the local model of the central model. The local model will be saved in the computer of the team member. As the team member works, the local server requests updated information from the central model on the central server, using available network capacity to transfer the data over the WAN. The updated version of the model is stored on the local server, so the updates are readily available when a team member requests them.

LINKING ANALYTICAL MODEL FOR ANALYSIS

In Autodesk Revit, the physical model represents the physical layout that consists of structural walls, beams, columns, footings, and so on. The analytical model is created along with the physical model. The analytical model is the engineering description of a structural physical model.

The following elements have analytical model associated with them: Structural Columns, Structural Framing elements (such as beams and braces), Structural Floors, Structural Footings, and Structural Walls. The analytical model is a 3D model that is used for structural analysis. In this model, you can add loads and material properties that are required for analysis. To analyze the analytical model, transfer it into a third-party analysis software. You can link the analytical model via application programming interface (API) to several leading industry applications for building analysis, such as ADAPT-Builder, RISA 3D, Fastrack, ROBOT Millennium, Sofistik, and others. After the analysis has been performed, the results of the analysis are returned to the building model (physical and analytical) in Revit. The results that are returned to Revit, dynamically update the building model and the documentation associated with it.

Autodesk Revit HELP

Autodesk Revit 2020 helps you to easily understand the tools and methods used in a project. In Autodesk Revit 2020, you can access online help documentation (Autodesk WikiHelp) as well as local (offline) help documentation.

To access the help feature, click on the **Help** down arrow on the right of the **InfoCenter**; a flyout with various help options will be displayed. The options to access the help are discussed next.

Using the Revit 2020 Help

To access the local Revit 2020 help, choose the **Help** tool from **InfoCenter**; the **Autodesk Revit 2020** page will be displayed, as shown in Figure 1-22. You can also display the **Autodesk Revit 2020** page by pressing the F1 key. In this page, different areas such as **Learn about Revit**, **Resources**, and others are displayed. You can click on the required link from these areas to get the related information. In the **Learn about Revit** area, various help options related to Autodesk Revit are available. You can click on the required option to display the help page corresponding to the option. The **Resources** area contains various learning resources. You can click on the desired option in this area to get the information related to it.

***Figure 1-22** The local **Revit Help** page*

Chapter 2

Getting Started with a Structural Project

Learning Objectives

After completing this chapter, you will be able to:

- *Start a new structural project*
- *Set the units of various measurement parameters for a project*
- *Save a project*
- *Close a project and exit Revit 2020*

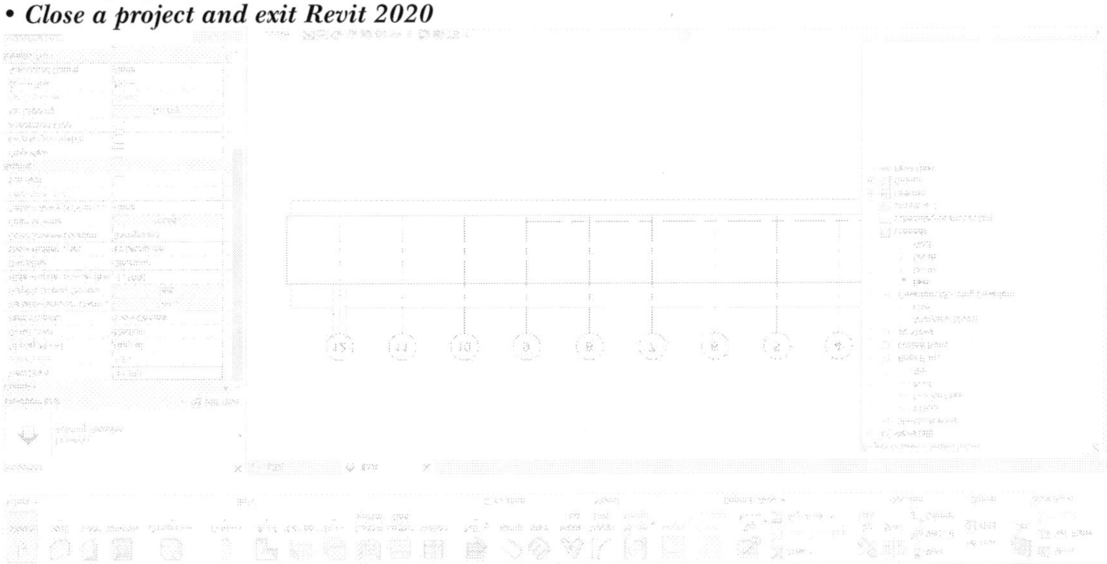

INTRODUCTION

In this chapter, you will learn how to create a new structural project. Also, you will learn the methods of creating initial setups such as units and snaps.

STARTING A NEW STRUCTURAL PROJECT

File Menu:	New > Project
Shortcut Key:	CTRL+N

In Revit, a structural project is considered as a single database containing information related to a building design. This means that the project file created will contain all the information related to building design, from geometry to construction documentation. Generally, this information include the details of the components used to design building models (like walls, doors, windows, beams, columns, and other relevant information about the building design), different views of the model, working drawings created from model, and documentation related to design of the model.

In Revit, you can create different views such as plan, elevation, and sections from a single model created for a building project. These views are associated with each other. This means, when you change the building design in one view, it is propagated throughout the project.

In Revit, a new project file uses the default settings that are defined in the project template.

To start a new project, choose the **New > Project** from the **File** menu, as shown in Figure 2-1; the **New Project** dialog box will be displayed, refer to Figure 2-2. In this dialog box, the **Construction Template** is selected by default in the **Template file** drop-down list. As a result, the new project will adopt the settings of the template file. A template file has various project parameters saved in it such as units, views, and so on. When you apply the template file on a new project, it will adopt the same parameters as the template file. The difference between a template file format and a project file format is that the former has a *.rte* extension, whereas the latter has a *.rvt* extension. You can either select any of the template files provided in Revit or create your own template file.

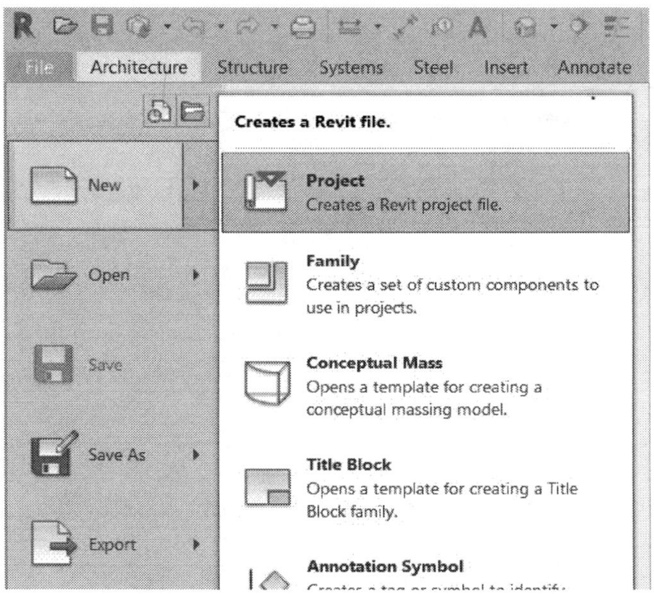

Figure 2-1 *Choosing the **Project** option from **File** Menu*

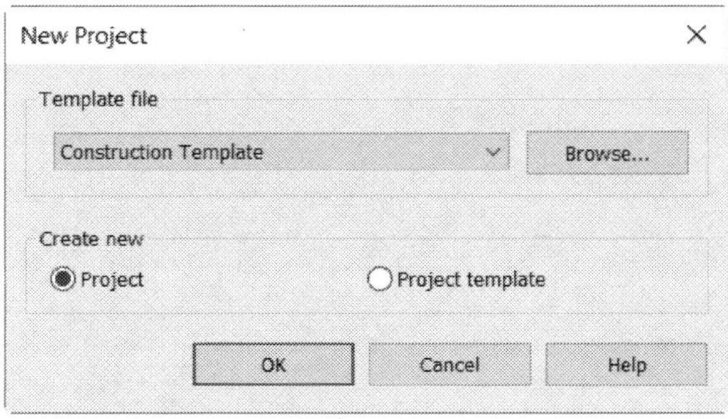

***Figure 2-2** The **New Project** dialog box*

You can also save any project file as a template file. You can select a file as template by first choosing the **Browse** button in the **Template file** area of the **New Project** dialog box and then selecting the required file from appropriate location. To create a new project, select the **Project** radio button and to create a template, select the **Project template** radio button in the **New Project** dialog box.

SETTING PROJECT UNITS

Ribbon:	Manage > Settings > Project Units
Shortcut Key:	UN

Units are important parameters of a project as they provide a standard for measurement of different entities. While installing Revit, you are prompted to set the Imperial (feet and inches) or Metric (meter) unit as default. Setting a default unit system helps you to start your project with a specific type of unit as per the conventions followed in the industry. To set units, choose the **Project Units** tool from the **Settings** panel of the **Manage** tab; the **Project Units** dialog box will be displayed, as shown in Figure 2-3. Alternatively, type **UN** to display the **Project Units** dialog box. Under the **Units** column in this dialog box, you can specify various units that are relevant to the building project. In the **Project Units** dialog box, units are grouped into six disciplines: **Common**, **Structural**, **HVAC**, **Electrical**, **Piping**, and **Energy**. Each discipline has a set of measurement parameters. You can select any of these disciplines from the **Discipline** drop-down list in the **Project Units** dialog box. The **Format** column in this dialog box displays the current unit format for the corresponding parameter in the **Units** column. You can preview and select the possible digit grouping and decimal separators from the **Decimal symbol/digit grouping** drop-down list, which is at the lower left corner of the dialog box. Different types of units and their parameters are discussed next.

Note

*The values displayed in the **Format** column of the **Project Units** dialog box may differ, depending upon the type of unit system, Imperial or Metric, selected for the project. In this textbook, the Imperial and Metric unit systems have been used in tutorials and illustrations.*

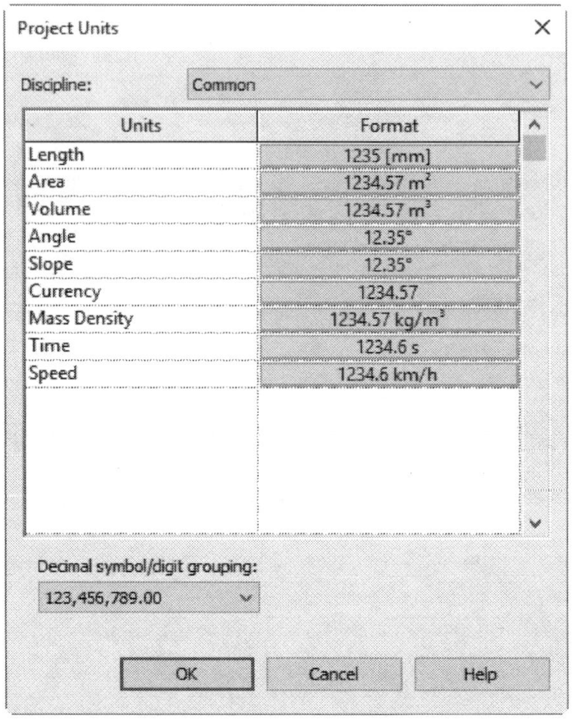

Figure 2-3 *The **Project Units** dialog box*

Common Unit Type

The **Common** unit type used in a structural project includes the parameters such as length, volume, area, angle, slope, mass density, currency, time and speed. In the **Project Units** dialog box, the **Common** option is selected by default in the **Discipline** drop-down list, refer to Figure 2-3. The **Common** unit type used in Revit is similar to those used in other Revit platforms. Moreover, the settings of the parameters of common units are similar to those used in other CAD programs.

Structural Unit Type

Structural units include the settings such as Force, Moment, Stress, Unit Weight, and so on. In Revit, you can set the structural units in the **Project Units** dialog box. To do so, invoke the **Project Units** dialog box and then select the **Structural** option from the **Discipline** drop-down list. Figure 2-4 displays the **Project Units** dialog box with the **Structural** option selected in the **Discipline** drop-down list. Some of the structural units are discussed next.

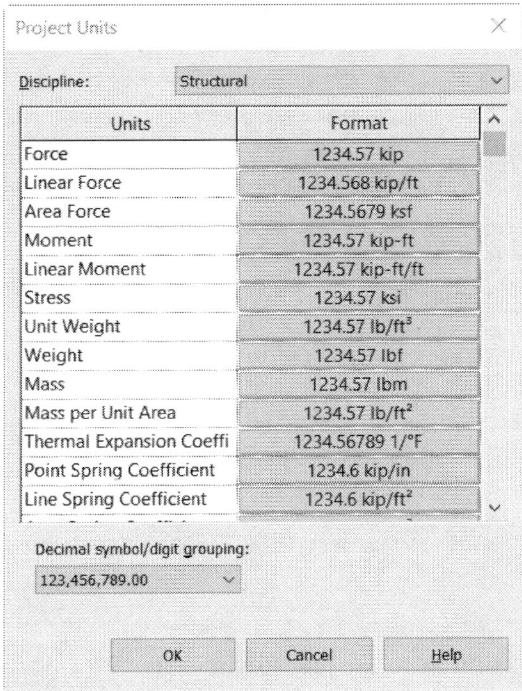

***Figure 2-4** The **Project Units** dialog box with the **Structural** option selected from the **Discipline** drop-down list*

Setting the Unit for Force

To specify the unit for the concentrated forces that act on the structural model, click on the field corresponding to the **Force** parameter in the **Format** column; the **Format** dialog box will be displayed. In this dialog box, specify a unit by selecting the required option from the **Units** drop-down list. The **Units** drop-down list contains the options such as **Newtons**, **Decanewtons**, **Kilonewtons**, **Meganewtons**, **Kips**, **Kilograms force**, and so on. For the Imperial unit system, **Kips** is the default option selected in the drop-down list. For the Metric unit system, **Kilonewtons** is the default option selected.

Setting the Unit for Linear Force

The linear force implies a uniformly distributed load that you can apply to any structural member. To specify unit for the linear force, click on the field corresponding to the **Linear Force** parameter in the **Format** column of the **Project Units** dialog box; the **Format** dialog box will be displayed. In this dialog box, specify the desired unit for the linear force by selecting an option from the **Units** drop-down list. The **Units** drop-down list contains options such as **Newtons per meter**, **Decanewtons per meter, Kilonewtons per meter**, **Meganewtons per meter**, **Kips per foot**, **Kilograms force per meter**, and so on. The default unit selected for the **Imperial** unit system in is **Kips per foot**. The default unit selected for the **Metric** unit system in the drop-down list is **Kilonewtons per meter**.

Tip

*While selecting a rounding value from the **Rounding** drop-down list in the **Format** dialog box, you should consider the extent of detailing that may be required for the project, and then select it. For projects that require too much detailing, a lower rounding value may be set. This parameter, however, can be modified at any time during the project development.*

Setting the Unit for Area Force

Area forces denote the pressure forces that you can apply to any structural element (beams, columns, slabs, and so on). To specify unit for the area force, click on the field corresponding to the **Area Force** parameter in the **Format** column of the **Project Units** dialog box; the **Format** dialog box will be displayed. In this dialog box, specify the desired unit for the area force by selecting an option from the **Units** drop-down list. In the **Units** drop-down list, you can select various options such as **Newtons per square meter**, **Decanewtons per square meter**, **Kilonewtons per square meter**, **Meganewtons per square meter**, **Kips per square foot**, **Kilograms force per square meter**, and so on. The default unit selected for the Imperial unit system in the drop-down list is **Kips per square foot** and for the Metric unit system is **Kilonewtons per square meter.**

Setting the Unit for Mass per Unit Area

The **Mass per Unit Area** parameter specifies the density that you can apply on any structural elements such as rebars, structural steel, and so on. To specify a unit for the mass per unit area, click on the field corresponding to the **Mass per Unit Area** parameter in the **Format** column of the **Project Units** dialog box; the **Format** dialog box will be displayed. In this dialog box, specify the desired unit for the mass per unit area by selecting an option from the **Units** drop-down list. In the **Units** drop-down list, you can select options such as **Kilograms per square meter** and **Pounds per square foot**. The default unit selected in the Imperial unit system for this parameter is **Pound per square foot** and in the Metric unit system is **Kilograms per square meter**.

Similarly, you can set the units for the other options of different discipline types.

SNAPS TOOL

Ribbon: Manage > Settings > Snaps

The **Snaps** tool is one of the important tools used for creating and editing elements in a structural model. This tool is used to make the cursor snap or jump depending on the preset increments or specific object properties of elements such as endpoint, midpoint, and so on. Invoke the **Snaps** tool from the **Settings** panel of the **Manage** tab; the **Snaps** dialog box will be displayed, as shown in Figure 2-5. This dialog box has three areas, **Dimension Snaps**, **Object Snaps**, and **Temporary Overrides**.

Note

The settings specified in this dialog box will be applied to all the projects opened in the session and will not be saved.

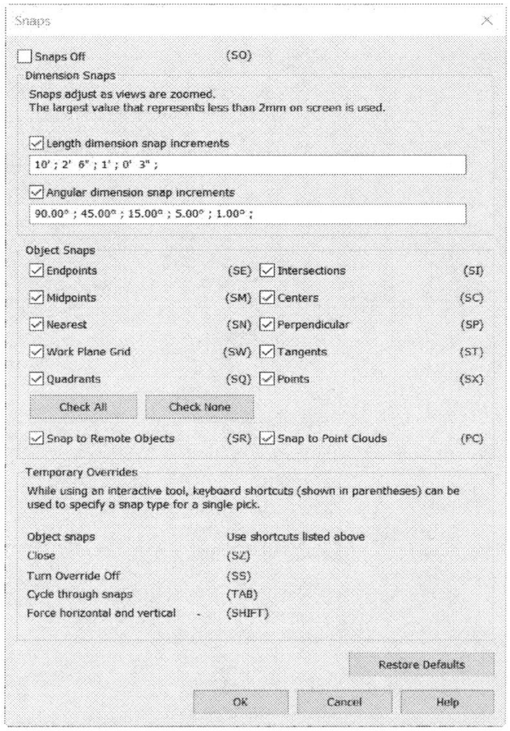

*Figure 2-5 The **Snaps** dialog box*

Tip

*The **Snaps** tool is used frequently not only while creating various building elements but also while editing and placing them. Besides making the modeling simpler, by using this feature efficiently, you can also improve the performance and accuracy of your project.*

OPTIONS DIALOG BOX

In Autodesk Revit, you can configure global settings by using the **Options** dialog box. This dialog box can be invoked by choosing the **Options** button from the **File Menu**. The **Options** dialog box, as shown in Figure 2-6, contains ten tabs: **General**, **Graphics**, **Hardware, File Locations**, **Rendering**, **Check Spelling**, **SteeringWheels**, **ViewCube**, **User Interface**, and **Macros**.

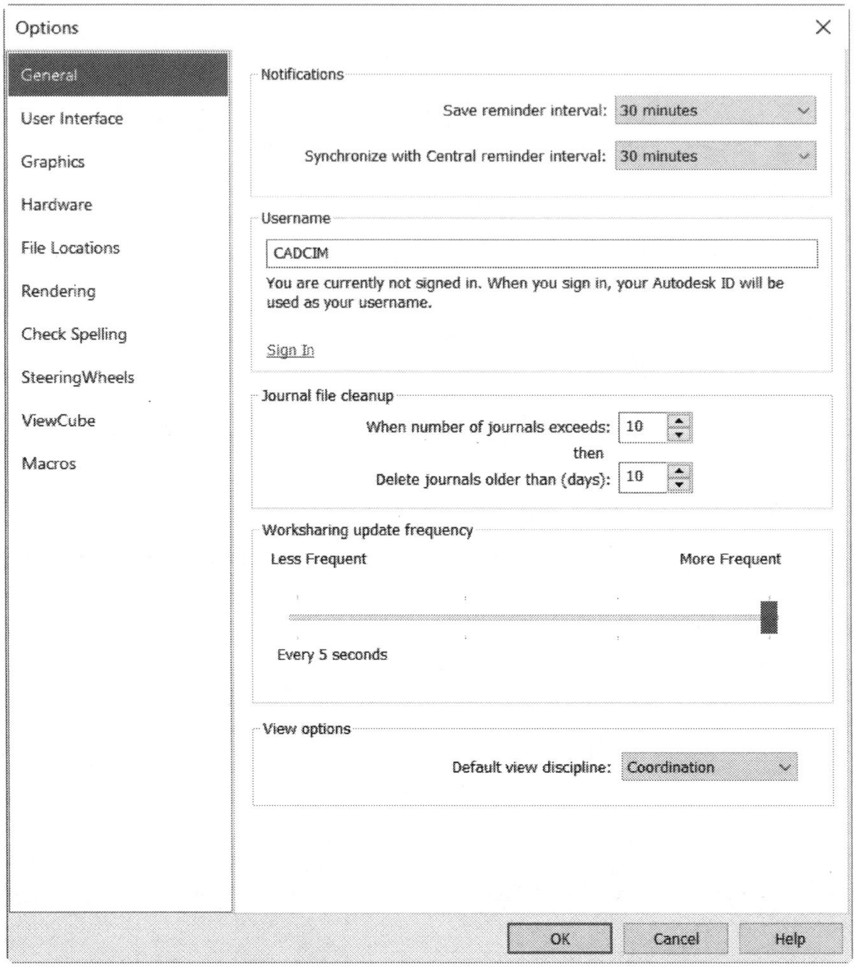

***Figure 2-6** Setting the file options using the **Options** dialog box*

SAVING A PROJECT

Before you close or exit a Revit session, it is recommended to save the project file. You can save a project file in a permanent storage device, such as a hard disk or a removable storage device like CD or USB. Also, you must save your work at regular intervals to avoid data loss due to any error in the computer's hardware or software.

Using the Save As Tool

Application Menu: Save As > Project

In Revit, you can save your project file at the desired location. To do so, choose the **File** menu and then choose **Save As > Project**; the **Save As** dialog box will be displayed, refer to Figure 2-7. Alternatively, you can save the project file by choosing the **Save** button from the **Quick Access Toolbar.**

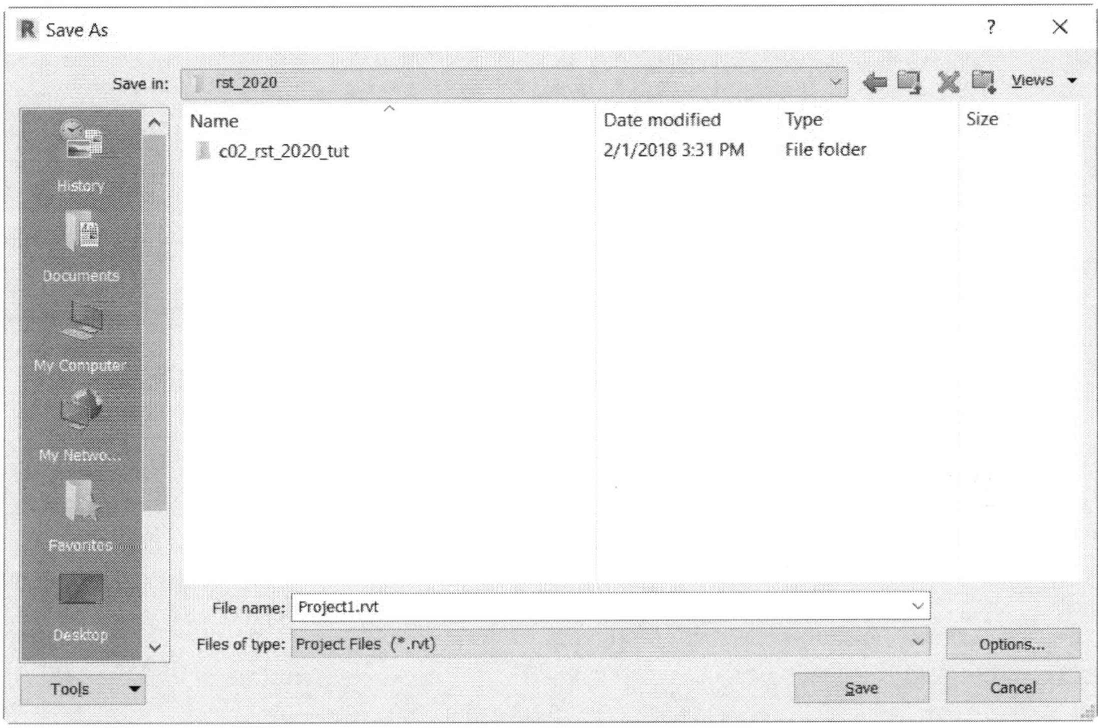

*Figure 2-7 The **Save As** dialog box*

Using the Save Tool

Once a project has been saved using the **Save As** tool, you do not need to re-enter file parameters to save it again. To save a project to the hard disk, choose the **File** menu button and then choose the **Save** option from **File Menu**. If you are saving the project for the first time, the **Save As** dialog box will be displayed, even if you invoke the **Save** tool.

> **Tip**
> *By default, the preview of a file is the last active view or sheet at the time it was last saved. You can set the preview to a particular view using the **Options** button in the **Save As** dialog box.*

Alternatively, you can save your project by choosing the **Save** button from the **Quick Access Toolbar**. As you save your project file, Revit updates it automatically without prompting you to re-enter the file name and path.

CLOSING A STRUCTURAL PROJECT

To close a project, choose the **File** menu button and then select the **Close** option. If you have already saved the latest changes, the project file will be closed. Otherwise, Revit will prompt you to save the changes through the **Save File** dialog box. You can save the changes by choosing the **Yes** button or discard them by choosing the **No** button. You can also choose the **Cancel** button to return to the interface and continue working on the project file. You can also use the **Close** button (X) in the drawing area to close the project.

EXITING A STRUCTURAL PROJECT

To exit a Revit session, choose the **Exit Revit** button from **File** menu. Even if the project is open, you can choose the **Exit Revit** button to close the file and exit Revit. If the project has not been saved once, the **Save File** dialog box will be displayed on choosing the **Exit Revit** button. In this dialog box, if you choose the **No** button, all unsaved changes will be lost. You can also use the **Close** button (**X**) in the main Revit window (in the title bar) to end the Revit session.

OPENING AN EXISTING STRUCTURAL PROJECT

Application Menu:	Open > Project
Shortcut Keys:	CTRL+O

In Autodesk Revit, there are several options available to open an existing project. These options are discussed next.

Opening an Existing Project Using the Open Tool

To open an existing project file, choose **Open > Project** from **File Menu**, as shown in Figure 2-8. Alternatively, you can open the project file by choosing the **Open** button from the **Quick Access Toolbar** or by pressing the CTRL+O keys. On invoking the **Open** tool, the **Open** dialog box will be displayed, as shown in Figure 2-9. Using the **Look in** drop-down list in this dialog box, you can access the desired folder and open the desired file.

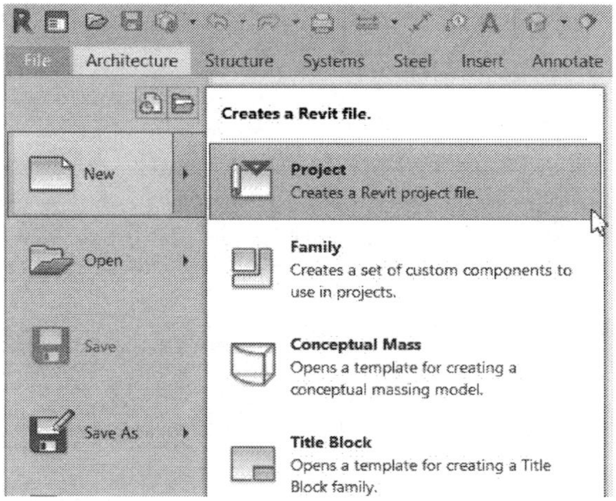

*Figure 2-8 Choosing the **Project** option from the **File** menu*

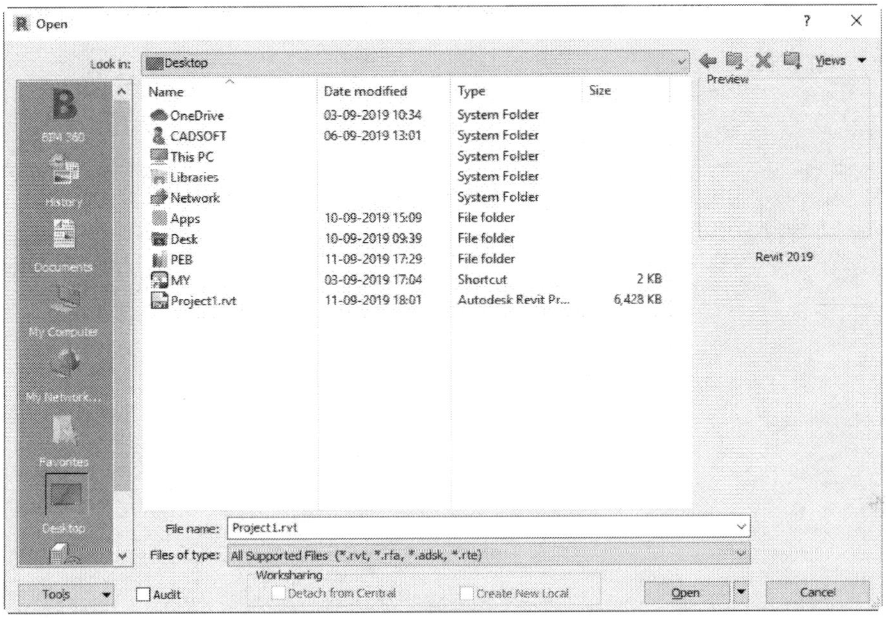

*Figure 2-9 The **Open** dialog box*

TUTORIAL

In this tutorials, you will create two structural projects as tutorials: a commercial project of a concrete structure and an industrial project of steel and concrete structure. These tutorials will be created in a sequence according to the topics discussed in the chapter.

Tutorial 1 Commercial Complex

In this tutorial, you will create a revit file for the *Commercial Complex* project with the following parameters and project specifications: **(Expected time: 15 min)**

1. Template file:
   For Imperial **US Imperial > Structural Analysis-Default.rte**
   For Metric **US Metric > Structural Analysis-DefaultMetric.rte**

2. Project Units:
   For Imperial **Feet and fractional inches**, Rounding- **To the nearest 1/2"**
   For Metric **Millimeters**, Rounding- **To the nearest 10**

3. Length dimension snap increment:
   For Imperial **5' ; 2' 6" ; 0' 3" ; 0' 01/2"**
   For Metric **1500 ; 700 ; 70 ; 10**

4. File name to be assigned:
   For Imperial *c02_Commercial-Complex_tut1.rvt*
   For Metric *M_c02_Commercial-Complex_tut1.rvt*

The following steps are required to complete this tutorial:

a. Start a Revit session.

b. Open a new project.

c. Use the template file for the project,
For Imperial **Structural Analysis-Default.rte**.
For Metric **Structural Analysis-DefaultMetric.rte**

d. Select project units.

e. Set **Endpoint**, **Midpoint**, **Nearest**, **Perpendicular**, **Work Plane Grid**, and **Intersections** as the object snaps in the **Snaps** dialog box for both the systems (Imperial & Metric).

f. Save the project by using the **Save** tool.
For Imperial *c02_Commercial-Complex_tut1.rvt*
For Metric *M_c02_Commercial-Complex_tut1.rvt*

g. Close the project by using the **Close** tool.

Starting Autodesk Revit 2020

1. Start Autodesk Revit by choosing **Start > All Programs > Autodesk > Revit 2020**. On doing so, the Revit interface window is displayed.

Opening a New Project

1. Choose **New > Project** from menu; the **New Project** dialog box is displayed.

Selecting the Template File

To select the template file for the project, you need to access the appropriate folder and select the required template file from it.

1. In the **New Project** dialog box, choose the **Browse** button from the **Template file** area; the **Choose Template** dialog box is displayed. This dialog box shows a list of template files available in the **US Imperial** folder, if **Imperial** unit system is selected at the time of Revit installation. (On selecting **Metric** unit system, list of metric template is displayed).

2. In the **Choose Template** dialog box, the template file **Structural Analysis-Default.rte**, as shown in Figure 2-10, is selected as the default template file in the **US Imperial** folder. In **US Metric** folder, default template file is **Structural Analysis-DefaultMetric.rte**. Choose the **Open** button to assign the selected template file to the **New Project** dialog box.

3. Select the **Project** radio button (default) to create a project file and then choose the **OK** button; the selected template file of selected unit system is loaded. Note that the **Project Browser** shows different levels and views that have already been created in the selected template file.

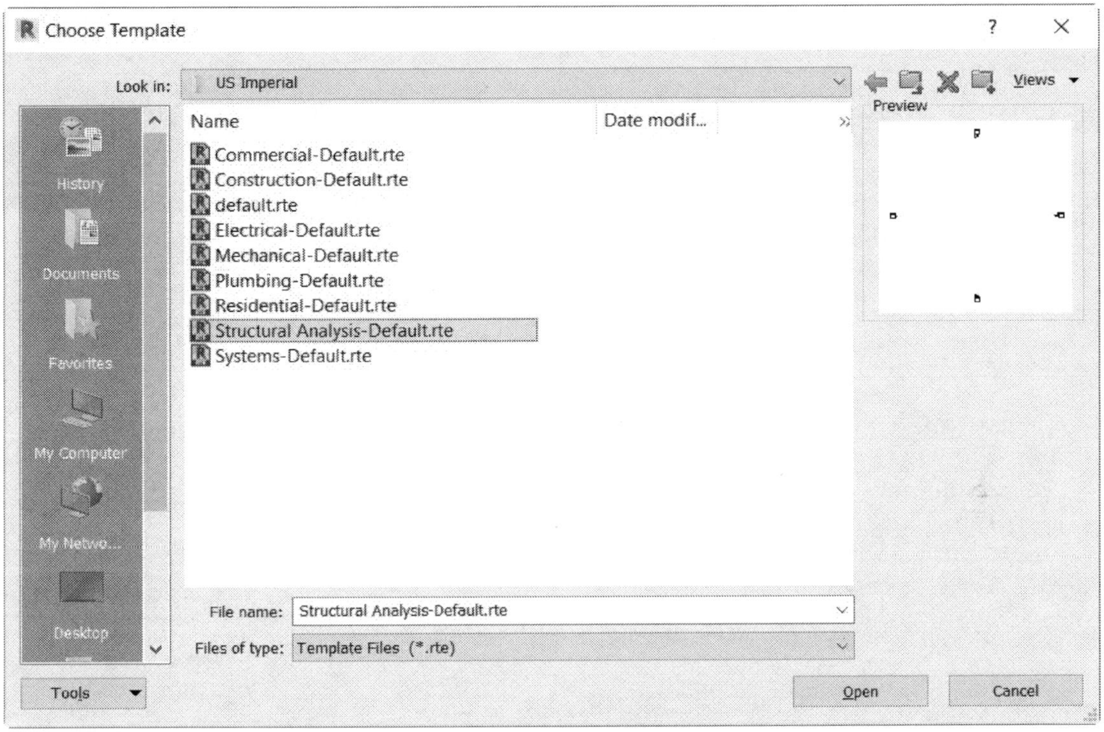

*Figure 2-10 The **Choose Template** dialog box*

Setting Project Units

1. To set units for the project, choose the **Project Units** tool from the **Settings** panel of the **Manage** tab; the **Project Units** dialog box is displayed.

2. Make sure that the **Common** option is selected in the **Discipline** drop-down list of the **Project Units** dialog box. Now, click in the **Format** column corresponding to the **Length** unit; the **Format** dialog box is displayed.

3. In this dialog box, select the option by clicking on the **Units** drop-down list.
   For Imperial **Feet and fractional inches**
   For Metric **Millimeters**

4. Select the option from the **Rounding** drop-down list.
   For Imperial **To the nearest 1/2"**
   For Metric **To the nearest 10**

5. Choose the **OK** button in the **Format** dialog box and then in the **Project Units** dialog box to close them.

Setting Dimension and Object Snaps

You need to set dimension snaps and object snaps by using the **Snaps** tool. These settings are based on the type of the project and the amount of detailing required.

1. Choose the **Snaps** tool from the **Settings** panel of the **Manage** tab; the **Snaps** dialog box is displayed, as shown in Figure 2-11.

2. In the **Dimension Snaps** area of this dialog box, ensure that the **Length dimension snap increments** check box is selected.

3. Now, click in the edit box below the **Length dimension snap increments** check box and delete the existing values and then enter the new values.

   For Imperial **5' ; 2' 6" ; 0' 3" ; 0' 0 1/2"**
   For Metric **1500 ; 700 ; 70 ; 10**

   Figure 2-11 shows the value entered for length snaps increment.

4. In the **Object Snaps** area, clear the **Quadrants**, **Centers**, **Tangents**, and **Points** check boxes to disable them. Leave the other options such as **Endpoints**, **Midpoints**, **Nearest**, **Perpendicular**, **Work Plane Grid**, **Snap to Remote Objects**, **Snap to Point Clouds**, **Perpendicular** and **Intersections** enabled, refer to Figure 2-11. Choose the **OK** button to apply settings and to close the **Snaps** dialog box.

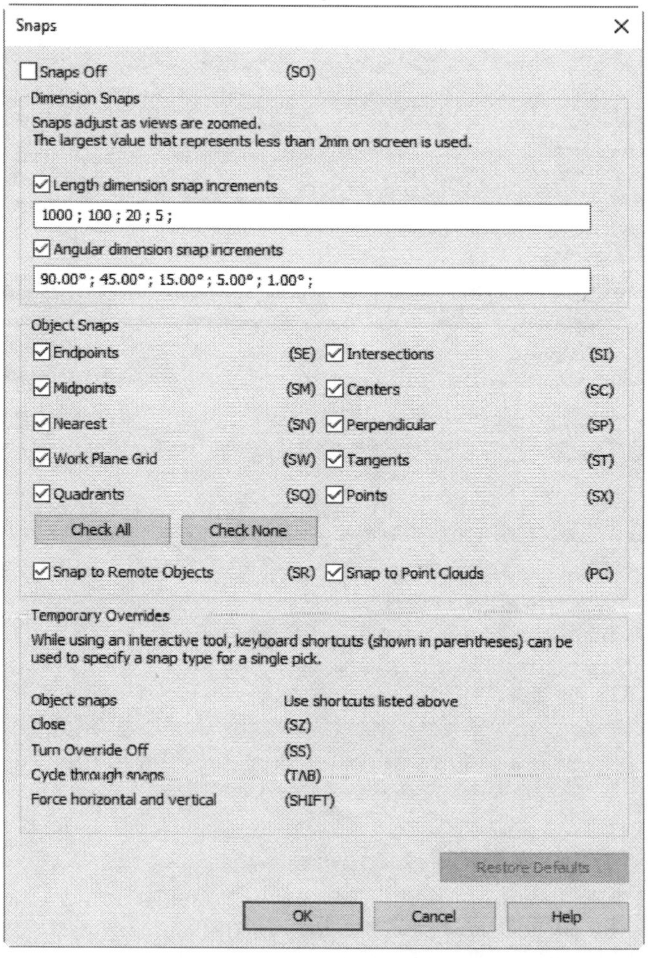

***Figure 2-11** The **Snaps** dialog box*

Saving the Project

The project parameters set in the previous steps are the global settings of the project file. To save this project file, use the **Save As** tool.

1. Choose the **Save** option from the **File** menu. As you are saving the project for the first time, the **Save As** dialog box is displayed.

2. In this dialog box, browse to the *C* drive and then create a folder with the name **rst_2020**.

3. Open the *rst_2020* folder and then create a sub-folder with the name *c02_rst_2020_tut*. Next, open the created folder and save the file with the name *c02_Commercial-Complex_tut1.rvt* (For Imperial), refer to Figure 2-12. For Metric, save the file with the name *M_c02_Commercial-Complex_tut1.rvt*.

4. Choose the **Save** button; the **Save As** dialog box closes and the project file is saved.

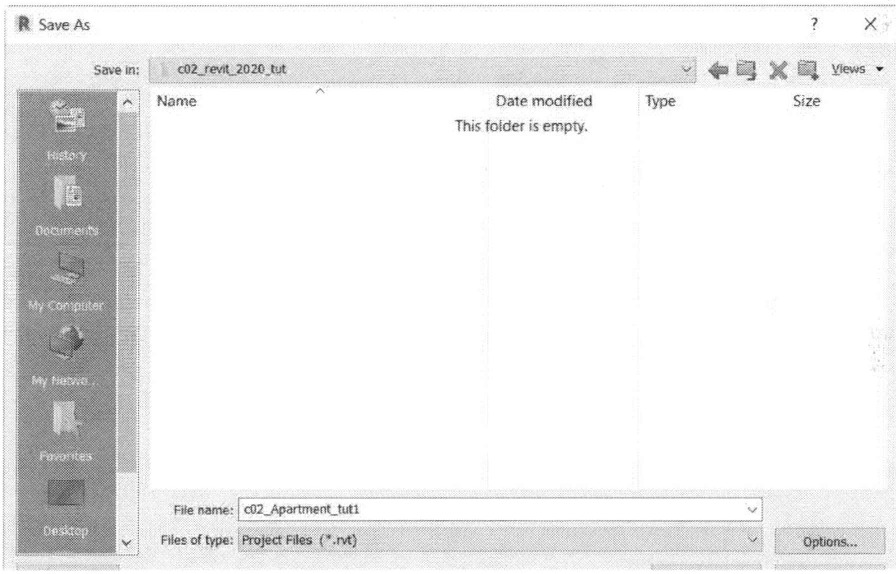

***Figure 2-12** The **Save As** dialog box*

Closing the Project

1. To close the project, choose the **Close** option from **File** menu in the Ribbon.

EXERCISE

Exercise 1 Academic Institution

In this exercise, you will create a new project file for the *Academic Institution* project with the following parameters. **(Expected time: 15 min)**

1. Template file-
   For Imperial **US Imperial > Structural-Analysis Default.rte**
   For Metric **US Metric > Structural-Analysis DefaultMetric.rte**

2. Project Units-

For Imperial	**Feet and fractional inches**
For Metric	**Millimeters**

3. File name to be assigned:

For Imperial	*c02_Academic-Institution_exer1.rvt*
For Metric	*M_c02_Academic-Institution_exer1.rvt*

Chapter 3

Setting up a Structural Project

Learning Objectives

After completing this chapter, you will be able to:

- *Create project templates*
- *Understand and create grids*
- *Understand and create levels*

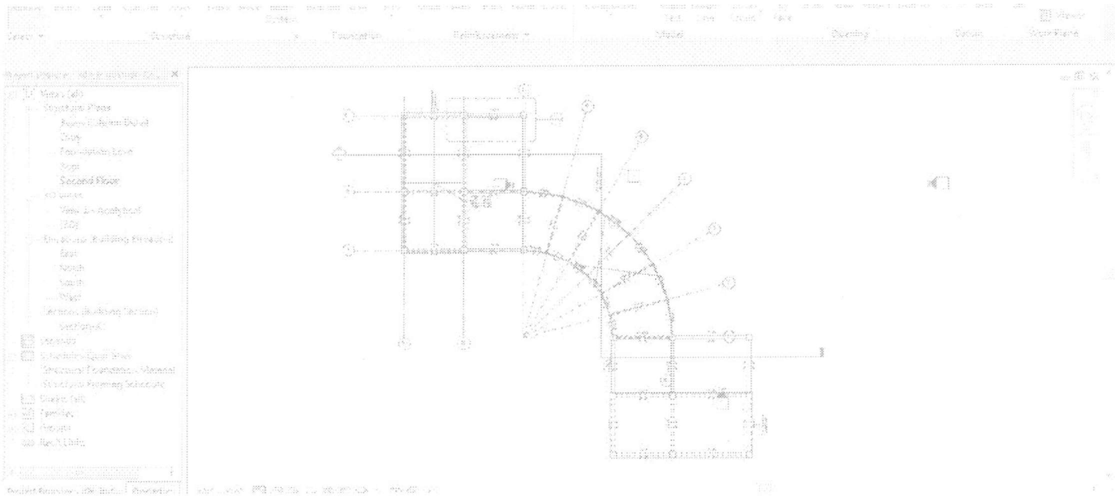

In the previous chapter, you learned to open a new file and create initial setups. Now, after opening a new file and creating initial setups such as units and snaps, you need to learn about creating advanced setups such as creating the project templates, levels, and grids for your project. These advance setups help you to start your project more efficiently and accurately.

In this chapter, you will learn to create a project template, save the project template and then create structural grids and levels for a structural project. Further, you will learn the use of Work Planes and Reference Planes in a project.

PROJECT TEMPLATE

Project templates are files that can be opened while starting a new project. These files, commonly known as template files, contain predefined settings for projects, display of annotations, display of graphics, and so on. Similarly, for the display of graphics, the template files contain predefined settings for materials, line styles, line weights, line patterns, and structural symbols.

Project templates provide initial conditions for a project. In Autodesk Revit, you will find in-built templates that are saved with *.rte* file extension when you install the software. You can also create your own template based on the project requirement. In Revit, any new template-based project inherits all families, settings (such as units, fill patterns, line styles, line weights, and view scales), and geometry from the template.

Template files carry project-specific information. You may have different project template files for different types of projects. For example, for a concrete structure, you may not use steel members for your design purpose. For this kind of project, you may require a template that carries families with more concrete sections and less steel members. A template file can also carry predefined levels and grids and thus saves a lot of time while starting up a new project. Schedules for footings, columns, piers, and other structural elements can also be a part of a project template.

In the next sections, you will learn how to create custom templates and use them in a structural project.

Creating a Custom Project Template

In Autodesk Revit, there are various methods to create a custom project template. The common method is to open an existing template file and modify its settings based on the project requirement and then save it as a different template file. You can also create a custom project template by starting a blank project file, defining all settings such as naming the viewports, creating levels, adding grids, and others, and then saving it as a template (.rte) file. For certain projects, you can create a template file which includes geometry that can be used repeatedly as a base for the new projects. For example, if you have defined geometry for a shopping complex and want to include this geometry whenever you start a new project, you can save the file that includes this geometry as a template. Each time you open a project with this template, the geometry is included.

Settings for the Project Template

While creating a project template, you can predefine certain settings based on your project requirement. To start a new project template, you need to fill in the information specific to the project. The information includes the name of the project, project number, client's name, and so on. These information are useful while publishing or plotting the drawing. Next, you need to enter the project settings. These settings include units, snaps, the line styles for components and lines, fill patterns for materials, and more. Some of the settings such as units and snaps have already been discussed in Chapter 2.

After modifying the project settings, you can create settings for families. The families in a project template can be system families and loaded families. While defining the settings for the project template, you can modify or duplicate system families (for example, walls) as required for the project. You can also load the commonly used families, user-defined families, and title blocks.

After setting families in the project template, you can modify or create settings for project views. These settings include plan views (Structural Plans), levels, schedules, sheets, and so on. There are other settings that can be made for the project template other than those discussed. These are visibility/graphics settings, and the plot (Print) settings. The settings for the project information are discussed in the next section.

USING LEVELS

Creating levels is one of the important tasks in a BIM project. In any structural model of a building project, levels define horizontal planes for placing the level-hosted elements, such as walls, roofs, foundations, floors, slabs, beams, columns, and ceilings.

In Revit, you can use the **Level** tool to add new levels by sketching the required level lines in the elevation or section views. By default, all level lines have associated labels that display the name and elevation of the level. There are two types of level that you can create: story and nonstory. The procedure to create these levels is discussed later in this chapter. A story level defines a floor, roof, or ceiling of a building. You can create a level for each known story in a structural model. The main characteristic of story levels is that they have corresponding plan views. In a project view, the level bubble for a story level will be displayed in blue.

On the other hand, a nonstory level does not hold any plan view but can act as host for placing objects and information. The examples of a nonstory level are top of wall, top of foundation, and top of slab. The level bubble for a nonstory level is displayed in black. Figure 3-1 shows both types of levels in the **East** elevation view of a structural model. In this figure, the **T.O. Column** and **Column Splice** levels are nonstory levels and the rest of the levels are story levels.

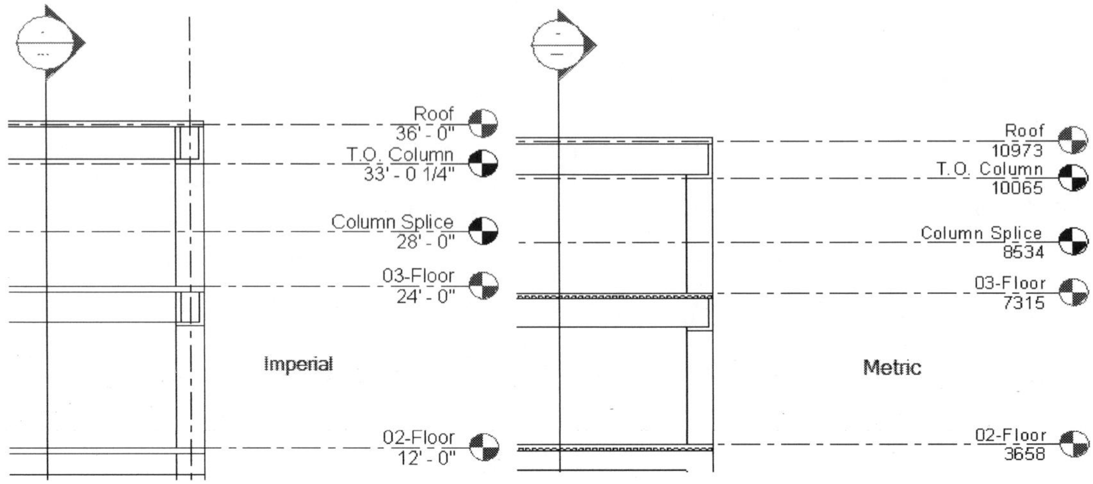

Figure 3-1 *Different types of level in the* ***East*** *elevation view*

In Autodesk Revit, when you use the default template file for creating a new project file, it loads two predefined levels: **Level 1** and **Level 2**. These levels are displayed in the elevation or section view. You can view any of the elevations or section views using the **Project Browser**. Levels can be added, renamed, and modified at any time during the project development. In the following sections, you will learn more about properties of levels, creating levels, modifying levels, and working with levels.

Controlling the Visibility of Levels

While working in an elevation view or a section view, sometimes you may need to hide some or all level lines to avoid congestion of construction lines. This helps in viewing the detail lines for sections and elevations properly. To do so, in the elevation view or section view, select the level line that you need to hide, and then right-click; a shortcut menu will be displayed. From this shortcut menu, choose **Hide in View > Elements**, as shown in Figure 3-2; the selected level line will be hidden.

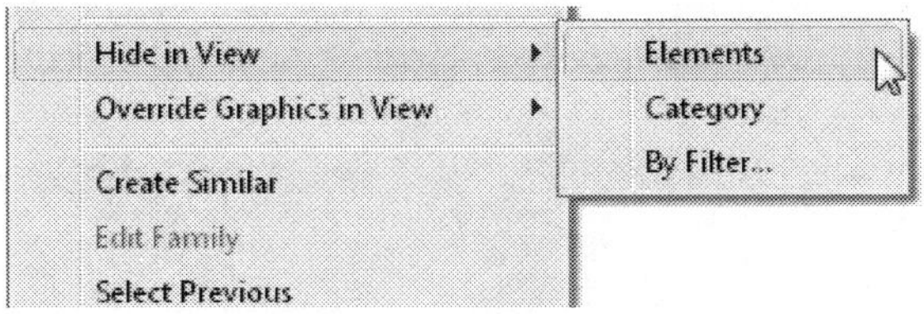

Figure 3-2 *Choosing the* ***Elements*** *option from the shortcut menu*

If you need the hidden level line to be visible, choose the **Reveal Hidden Elements** button from the **View Control Bar**; the hidden level line will appear red in the drawing area. Next, right-click on the level appearing in red color in the drawing area; a shortcut menu will be displayed. Choose **Unhide in View > Element** from the shortcut menu. Next, choose the **Close Reveal Hidden Elements** button from **View Control Bar**; the hidden level will appear.

Similar to hiding a single level line, you can also hide all levels that are displayed in an elevation or a section view. To do so, select any level from the view and right-click; a shortcut menu will be displayed. Choose **Hide in View > Category** from the shortcut menu; all levels in the view will hide.

GRIDS

Grids are the essential features that will help you to align and place structural elements such as columns, beams, and walls in a structural plan. With grids, you can easily trace the exact location of a desired column or any other structural elements. Grids are definite vertical planes represented as lineworks in the plan, elevation, and section views.

You can add grids as lines or arcs. You can also create them from existing lines, or link/import them from other software such as AutoCAD Architecture, Revit Architecture, and AutoCAD and then convert them to revit grids. In the following sections, you will learn how to create, modify, and understand the grids.

Creating Grids

Ribbon: Structure > Datum > Grid
Shortcut Key: GR

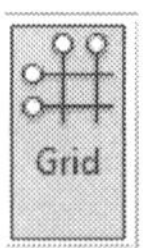

To create a grid, invoke the **Grid** tool from the **Datum** panel of the **Structure** tab; the **Modify | Place Grid** contextual tab will be displayed, as shown in Figure 3-3. The options in the **Modify | Place Grid** contextual tab can be used to assign and change the element properties (instance and type) of the grid, to select a type for grid, and to draw grid in the desired plan or elevation.

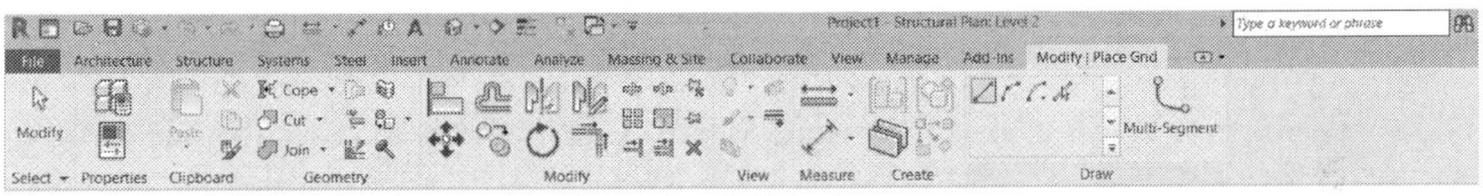

*Figure 3-3 The options in the **Modify | Place Grid** contextual tab*

Before you start sketching grids, you need to assign their element properties (instance or type) using the **Properties** palette. To do so, invoke the **Grid** tool; the **Properties** palette will display the parameters that can be added to the grid. In the palette, you will notice that the value assigned to the **Scope Box** parameter is **None**, which implies that no scope box will be assigned to the grid to be added. Next, in the palette, choose the **Edit Type** button; the **Type Properties** dialog box will be invoked. In this dialog box, you can change the type parameters of grids, as discussed next.

Sketching the Grids

After assigning the type parameters in the **Type Properties** dialog box, you can now start sketching the grid line. To sketch the grid line, invoke any of the sketching tools from the **Draw** panel of the **Modify | Place Grid** contextual tab. The **Draw** panel of the **Modify | Place Grid** contextual tab contains four tools: **Line**, **Start-End-Radius Arc**, **Center-ends Arc**, and **Pick Lines**, as shown in Figure 3-4. The use of these sketching tools is discussed next.

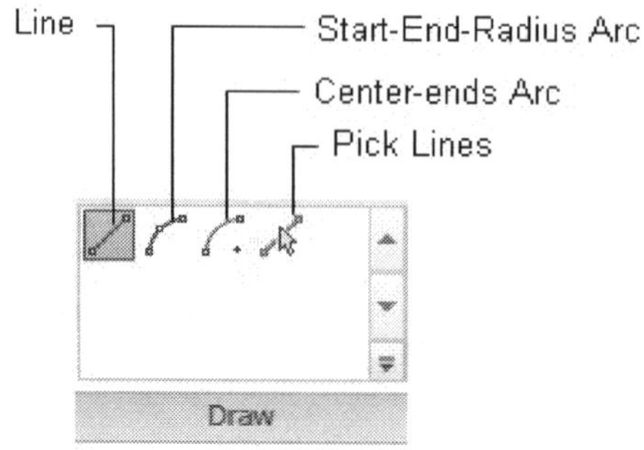

***Figure 3-4** The sketching tools in the **Draw** panel*

Controlling the Visibility of Grids

You can control the visibility of each grid or all grids in any of the project views. To do so, select a grid and right-click; a shortcut menu will be displayed. Choose **Hide in View > Category** from the shortcut menu; the grid category will be hidden in the current view. Similarly, you can also hide one particular grid by choosing **Hide in View > Elements** from the shortcut menu. On doing so, the selected grid will be hidden in the current view. However, the grid will be displayed in all the other views.

Alternatively, you can control the visibility of the grids by using the visibility/graphics feature of Autodesk Revit. To do so, select a grid from the drawing and choose the **Override by Category** tool from **Modify | Grids > View > Override Graphics in View > Override by Category** drop-down; the **View-Specific Category Graphics** dialog box will be displayed. Choose the **Open the Visibility Graphics dialog** button; the **Visibility/Graphic Overrides for Structural Plan: <view name>** dialog box for the current project view will be displayed. In the **Annotation Categories** tab of this dialog box, clear the check box for the visibility of grids; the visibility of all grids will turn off in the current view. You can also control the visibility of grids using the **Scope Box** feature.

TUTORIAL

Tutorial 1 — Commercial Complex

In this tutorial, you will create various project settings such as project information, project location, and browser organization for the project created in Tutorial 1 of Chapter 2. Further, you will add levels and grids to the project file. Use the following project parameters and specifications:

(Expected time: 1 hr 15 min)

1. Rename Level 1 and Level 2 as **Below Grade** and **Entry**, respectively.

2. Specify the elevation of **Below Grade** and **Entry** levels as follows.

   For Imperial **-4' 0"** and **0' 0"**

   For Metric **-1219** mm and **0** mm

3. Levels to be added:

For Imperial	**Second Floor - Elevation 12' 6"**
	Roof - Elevation 25' 0"
For Metric	**Second Floor - Elevation 3810 mm**
	Roof - Elevation 7620 mm

4. File name to be assigned:

For Imperial	*c03_Commercial-Complex_tut1.rvt*
For Metric	*M_c03_Commercial-Complex_tut1.rvt*

The following steps are required to complete this tutorial:

a. Open Tutorial 1 of Chapter 2.
b. Set the project information, project location, and browser organization.
c. Modify existing levels and add new levels.
d. Add grid lines.
e. Save the project using the **Save As** tool.
f. Close the project using the **Close** tool.

Opening an Existing Project

In this section, you will open the project file created in Tutorial 1 of Chapter 2.

1. To open the project file, choose **Open > Project** from the **File** menu; the **Open** dialog box is displayed.

2. In this dialog box, browse to *C:\rst_2020\c02_rst_2020_tut* folder and then choose the *c02_Commercial-Complex_tut1.rvt* (for Imperial) or *M_c02_Commercial-Complex_tut1.rvt* (for Metric) project file. You can also download this file from *https://www.cadsofttech.com*. The path of the file is as follows: *Textbooks > Civil/GIS > Revit Structure > Revit Structure 2020 for Novices.*

3. Choose the **Open** button from the **Open** dialog box to open the selected project file.

Setting the Project Information

In this section, you will add project information to the opened file. Further, you will define a parameter for the project information and add it to the project information of the project file.

1. To set the project information for the project, choose the **Project Information** tool from the **Settings** panel of the **Manage** tab; the **Project Information** dialog box is displayed.

2. In this dialog box, specify the project parameters as follows:

Parameter	**Value**
Project Issue Date	**01/01/2019**
Project Status	**Started**
Client Name	**Tickoo Institute of Emerging Technologies**

Project Address	**525, Andrews Drive Schererville**
Project Name	**Commercial Complex**
Project Number	**TIET/02/20**

3. In the **Project Information** dialog box, choose the **OK** button; the dialog box closes and the project parameters are saved in the project file.

4. Next, you need to create an additional instance parameter for the project information. To do so, choose the **Project Parameters** tool from the **Settings** panel of the **Manage** tab; the **Project Parameters** dialog box is displayed.

5. Choose the **Add** button from the **Project Parameters** dialog box; the **Parameter Properties** dialog box is displayed.

6. In the **Parameter Properties** dialog box, scroll the list in the **Categories** area and then select the **Project Information** check box.

7. In the **Parameter Data** area, ensure that the **Instance** radio button is selected. Now, enter the name **Project Engineer** in the **Name** edit box in the **Parameter Data** area.

8. Next, select the **Common** option from the **Discipline** drop-down list, if it is not selected by default.

9. Select the **Text** and **Other** options from the **Type of Parameter** and **Group parameter under** drop-down lists, respectively.

10. Choose the **OK** button in the **Parameter Properties** dialog box; the **Project Parameters** dialog box is displayed, as shown in Figure 3-5.

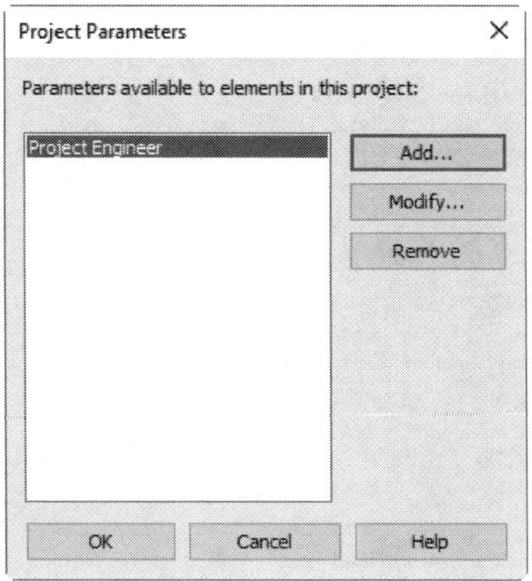

***Figure 3-5** The **Project Parameters** dialog box*

11. Choose the **OK** button; the **Project Parameters** dialog box is closed.

12. To enter the value for the new parameter, choose the **Project Information** tool from the **Settings** panel of the **Manage** tab; the **Project Information** dialog box is displayed.

13. In this dialog box, click in the **Value** field corresponding to the **Project Engineer** parameter and enter **Sham Tickoo**.

14. Next, choose the **OK** button; the **Project Information** dialog box closes and the value for the newly added parameter is saved in the project file.

Setting the Project Location

In this section, you will specify the geographical location of the project.

1. To specify the geographical location of the project, invoke the **Location** tool from the **Project Location** panel of the **Manage** tab; the **Location Weather and Site** dialog box is displayed.

2. In the **Location Weather and Site** dialog box, choose the **Site** tab and then the **Duplicate** button; the **Name** dialog box is displayed.

3. Enter **TIET-IN** in the **Name** edit box and then choose the **OK** button; the **Name** dialog box closes and the name entered is displayed and highlighted in the **Sites defined in this project** list box of the **Location Weather and Site** dialog box, refer to Figure 3-6.

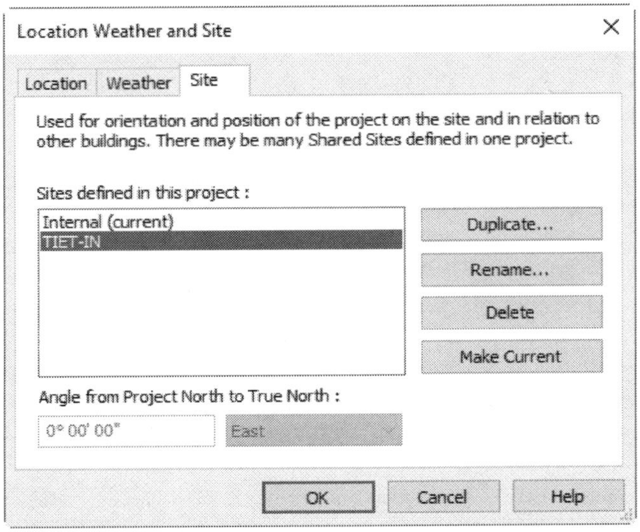

***Figure 3-6** TIET-IN selected in the **Sites defined in this project** list box of the **Site** tab*

4. Next, choose the **Location** tab in the **Location Weather and Site** dialog box; the options in this tab are displayed.

5. Select the **Default City List** option from the **Define Location by** drop-down list.

6. Next, select the **Indianapolis, IN** option from the **City** drop-down list.

7. Next, select the **Use Daylight Saving time** check box and then choose the **OK** button; the dialog box closes and the settings for the project location are saved in the project file.

Setting the Browser Organization

In this section, you will specify the settings for customizing the **Project Browser** displayed in the drawing area of the project.

1. To customize the **Project Browser**, choose the **Browser Organization** tool from **View > Windows > User Interface** drop-down; the **Browser Organization** dialog box is displayed.

2. Ensure that the **Views** tab is chosen, and then select the **Discipline** check box in this tab.

3. Next, choose the **Sheets** tab in the **Browser Organization** dialog box; the options in this tab are displayed.

4. In this dialog box, select the **Issue Date** check box.

5. Now to apply the modified settings, choose **Apply** and then **OK** from the **Browser Organization** dialog box; the modified settings are applied to the **Project Browser** and the **Browser Organization** dialog box is closed.

Modifying the Existing Levels

In this section, you will modify the existing levels in the project file. To do so, you need to select an elevation view from the **Project Browser** and then use the **Level** tool.

1. Before modifying the levels, move the cursor to the **Project Browser** and expand the Structural head and then the **Elevations (Building Elevation)** node. Now, double-click on the **North** node; the current view is changed to the North elevation view.

Note

*In the north elevation view, make sure that the **Hidden Line** option is chosen for the **Visual Style** in the **View Control Bar**.*

2. Next, zoom in the drawing area to make the text on the level lines visible. Click on the **Level 1** level line and move the cursor over the text **Level 1**; the text is highlighted.

3. Click on the text; an edit box is displayed.

4. Enter **Below Grade** in the edit box and press ENTER; the **Confirm Level Rename** message box is displayed. This message box prompts you to confirm whether or not to rename the corresponding view of the selected level.

5. Choose the **Yes** button; the name of the level changes in the **Project Browser**. Also, notice that the name of the view corresponding to the changed level has been changed under the **Structural Plans** head in the **Project Browser**.

6. Next, move the cursor over the elevation value of the **Below Grade** level; the elevation value is highlighted.

7. Click on the highlighted elevation value; an edit box is displayed.

8. Enter value in the edit box as given below:

   For Imperial **-4' 0"**
   For Metric **-1220 mm**

   Press ENTER and then press ESC to exit from the modification controls.

9. Repeat the procedure followed in steps 2 through 7 to rename the **Level 2** level to **Entry** and modify its elevation as given below:

   For Imperial **0'0"**.
   For Metric **0** mm.

10. Next, right-click in the drawing area; a shortcut menu is displayed.

11. Choose the **Zoom To Fit** option from the shortcut menu; the extents of the level lines are displayed in the drawing area.

12. Next, select the **Below Grade** level line and then place the cursor over the grid length control displayed on the left.

13. Press the left mouse button and drag the mouse to the right; the level lines move along the cursor.

14. Click in the drawing area when the length of the level line reduces to half of their existing length.

15. Press ESC to exit from the edit controls.

16. Next, right-click in the drawing area and then choose **Zoom To Fit** from the shortcut menu; the current view fits the extent of the length of level lines.

Adding New Levels

In this section, you will add new levels to the project file using the **Level** tool.

1. To add a new level, choose the **Level** tool from the **Datum** panel of the **Structure** tab; the **Modify | Place Level** contextual tab is displayed. Alternatively, you can press LL to activate the **Level** tool.

2. Choose the **Plan View Types** button from the **Options Bar**; the **Plan View Types** dialog box is displayed. In this dialog box, click and deselect **Floor Plan** and **Ceiling Plan**. Now, choose the **OK** button; the dialog box is closed.

3. Move the cursor near the left endpoint of the **Entry** level line, and when the alignment line appears, type value in the edit box displayed as given below:

   For Imperial **12'6"**, refer to Figure 3-7.
   For Metric **3810 mm**

   Press ENTER.

4. Next, move the cursor toward right until the alignment line appears above the **Entry** level bubble and then click to complete the level line, as shown in Figure 3-8. A new level line is created, displaying the level name as **Level 3** and the elevation as **12'6"** for imperial and for metric the elevation is **3810** .

Figure 3-7 Specifying the distance of the new level line

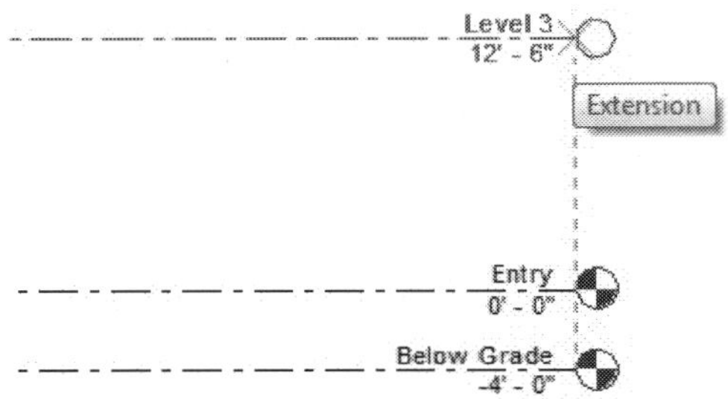

Figure 3-8 Creating the new level line using the alignment line

5. Without exiting the **Level** tool, repeat steps 2 and 3 to create a level line with the name **Level 4** at a distance from the **Level 3** level line as given below:

   For Imperial **12'6"**
   For Metric **3810**

   Note that the elevation of the **Level 4** level line is displayed as shown below:

   For Imperial **25' 0"**
   For Metric **7620**

6. Press ESC twice to exit from the **Level** tool.

7. Select the **Level 3** level line; the properties of the selected level line are displayed in the **Properties** palette.

8. In the **Properties** palette, click in the value field corresponding to the **Name** parameter and replace the existing name **Level 3** with **Second Floor**.

9. Choose the **Apply** button in this palette; the **Confirm Level Rename** message box is displayed.

10. In the **Confirm Level Rename** message box, choose the **Yes** button; the name of the level line **Level 3** changes to **Second Floor**. Also, notice that the name of the corresponding view in the **Project Browser** has changed to the new name.

11. Next, to rename the **Level 4** level view, select it from the drawing area; the properties of the selected level are displayed in the **Properties** palette.

12. Repeat steps 7 to 9 and rename **Level 4** to **Roof**.

13. Press ESC to exit from the tool.

14. Choose the **Level 1 - Analytical** option from the **Structural Plans** subhead in the **Project Browser**.

Note
*The **Structural Plans** head is a subhead in the **Structural** head of the **Views (Discipline)** hierarchy in the **Project Browser**.*

15. Right-click and then choose **Delete** from the displayed shortcut menu; the selected view name is deleted from the **Project Browser**.

16. Repeat steps 13 and 14 to delete the **Level 2- Analytical** option from the **Structural Plans** head.

Adding Grid Lines

In this section, you will add grids to the structural project. To add grids, you need to change the current view to the **Entry** level structural plan and then use the **Grid** tool.

1. Double-click on **Entry** from the **Structural Plans** head in the **Project Browser**; the **Entry** structural plan view will be displayed in the drawing area.

2. Invoke the **Grid** tool from the **Datum** panel of the **Structure** tab; the **Modify | Place Grid** contextual tab is displayed.

3. In this tab, ensure that the **Line** tool is invoked from the **Draw** panel. Next, move the cursor toward the lower left corner of the drawing area and then click to specify the start point of the first grid line; a grid line with temporary dimensions for angle emerges, refer to Figure 3-9 for location.

4. Move the cursor vertically up (at an angle of 90° from the horizontal plane) and click when it reaches near the **North** elevation arrow, refer to Figure 3-10 for location.

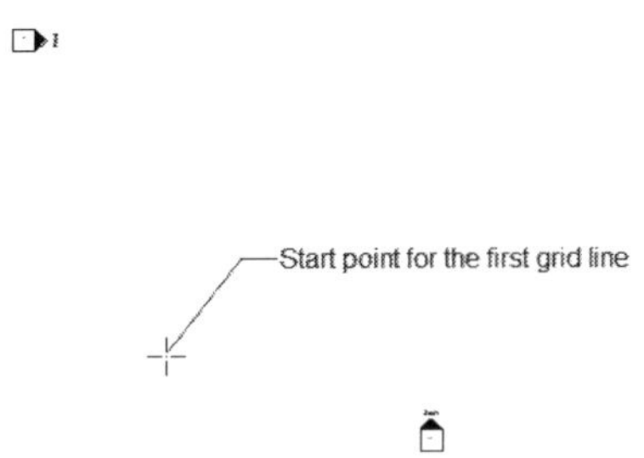

Figure 3-9 *Specifying the start point of the first grid line*

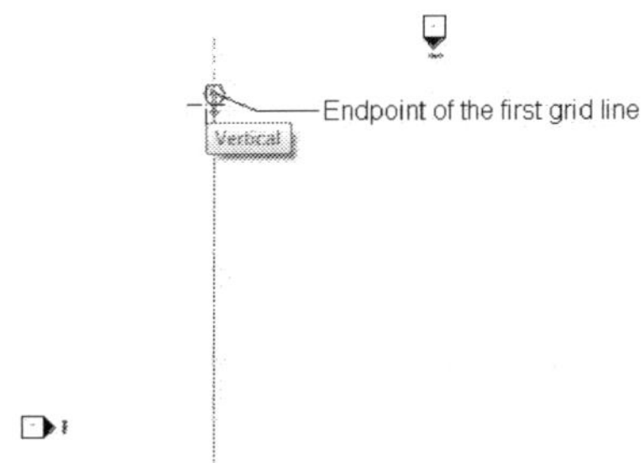

Figure 3-10 *Specifying the endpoint of the first grid line*

5. Now, move the cursor downward and position it near the right side of the endpoint of the grid line created previously, as shown in Figure 3-11.

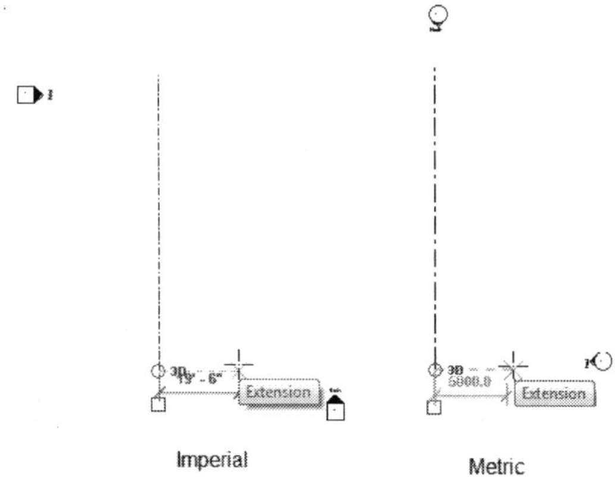

Figure 3-11 *Specifying the start point of the second vertical grid line*

6. As you move the cursor toward right, a temporary dimension appears with an extension line and the **Extension** snap. Enter value as given below:

   For Imperial **10'**
   For Metric **3048**

   Press ENTER; the second grid line starts at a distance of entered value from the start point of the first grid line.

7. Move the cursor vertically up and place it near the bubble of the first grid line and click when alignment line appears; second grid line is created.

8. Repeat steps 5 to 7 to create five more vertical grids with the distance specified in Figure 3-12.

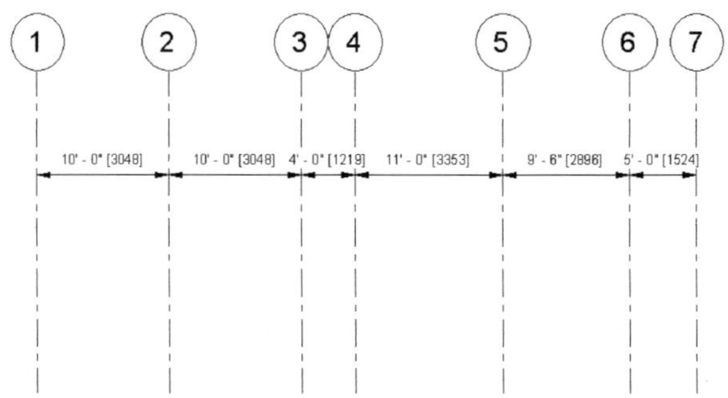

Figure 3-12 *Distance between the vertical grid lines*

9. After creating all the vertical grids, press ESC twice to exit the **Grid** tool.

10. Select the vertical grid line marked **4**. Now, place the cursor over the text in the grid bubble of the selected grid line and click; an edit box is displayed, as shown in Figure 3-13.

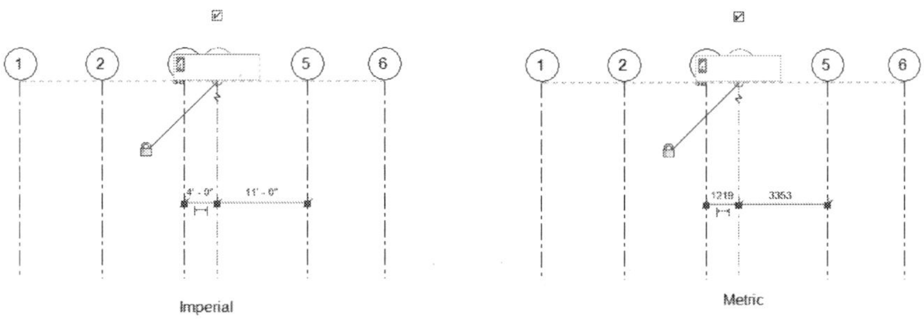

***Figure 3-13** Renaming the grid line using the edit box*

11. Enter **3.1** in the edit box and press ENTER; the selected grid line is renamed.

Note

*While renumbering the grid line, you can use the **Zoom in Region** and **Zoom Out(2x)** tools to view the grid numbers easily.*

12. Without exiting the editing mode, select the next grid marked **5** and then move the cursor over its grid bubble and click to display the edit box.

13. Enter **4** in the edit box and press ENTER to rename the selected line.

14. Repeat steps 10 and 11 to rename the grid lines marked **6** and **7** to **5** and **5.1**, respectively, as shown in Figure 3-14.

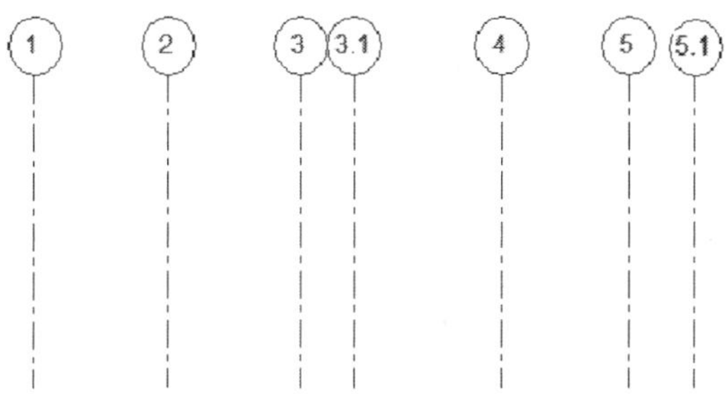

***Figure 3-14** The display of vertical grids after they are renumbered*

15. After renumbering all required vertical grids displayed in the drawing, press ESC to exit the editing mode.

16. To add horizontal grids, invoke the **Grid** tool from the **Datum** panel of the **Structure** tab.

17. Next, move the cursor toward the top right of the vertical grid line marked **5.1** and click at the location shown in Figure 3-15.

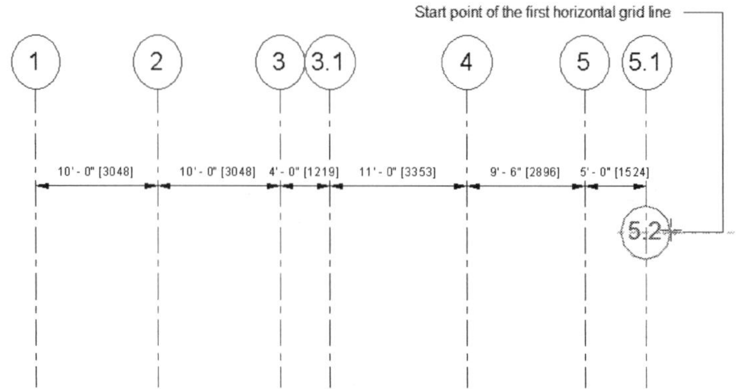

Figure 3-15 *The location of the start point of the first horizontal grid line*

18. Move the cursor horizontally toward the left until it has crossed the vertical grid line marked **1**. Click in the drawing area to create the first horizontal grid line, as shown in Figure 3-16.

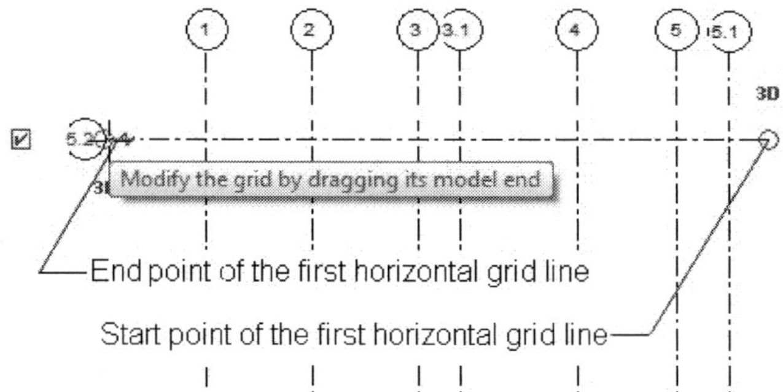

Figure 3-16 *The location of the start point and endpoint of the first horizontal grid line*

19. Move the cursor below the grid line alignment control of the start point of the first horizontal grid line (grid line created previously); an alignment line with a temporary dimension is displayed.

20. Type **8'** (or **2438 mm** for Metric); an edit box is displayed with the entered value. Press ENTER; the second horizontal grid line will start below the start point of the first horizontal grid line at a distance of **8'** (or **2438** mm in Metric) from it.

21. Move the cursor horizontally toward left and place it below the grid bubble (marked **5.2**) of the first horizontal grid line when the alignment line appears and then click to create the grid line.

22. Repeat steps 19 to 22 to create the rest of horizontal grid lines and rename them as previously done, refer to Figures 3-17 and 3-18.

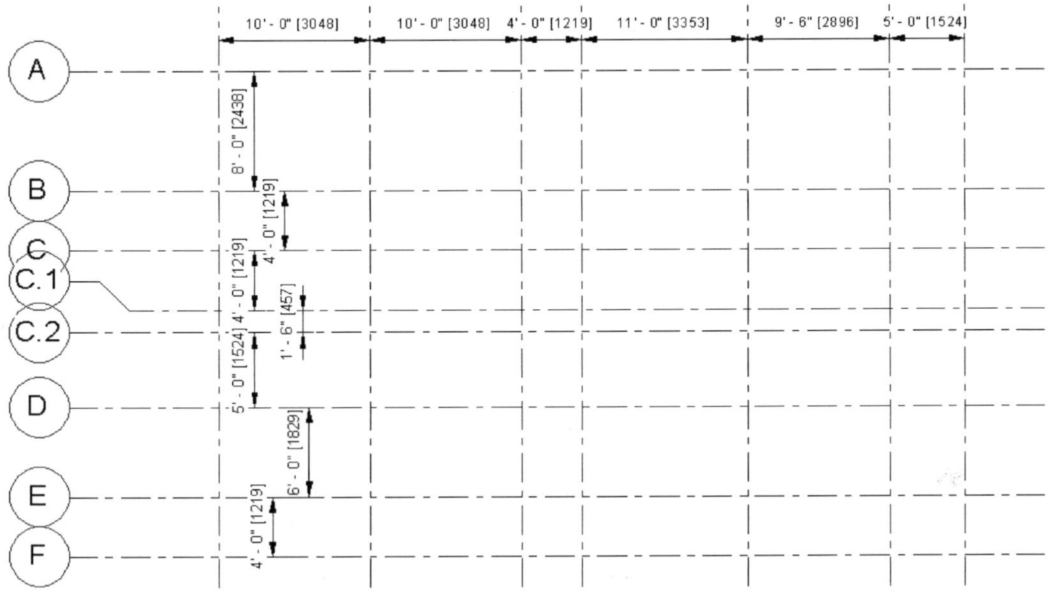

Figure 3-17 *The horizontal grid lines to be created with the specified names and distances*

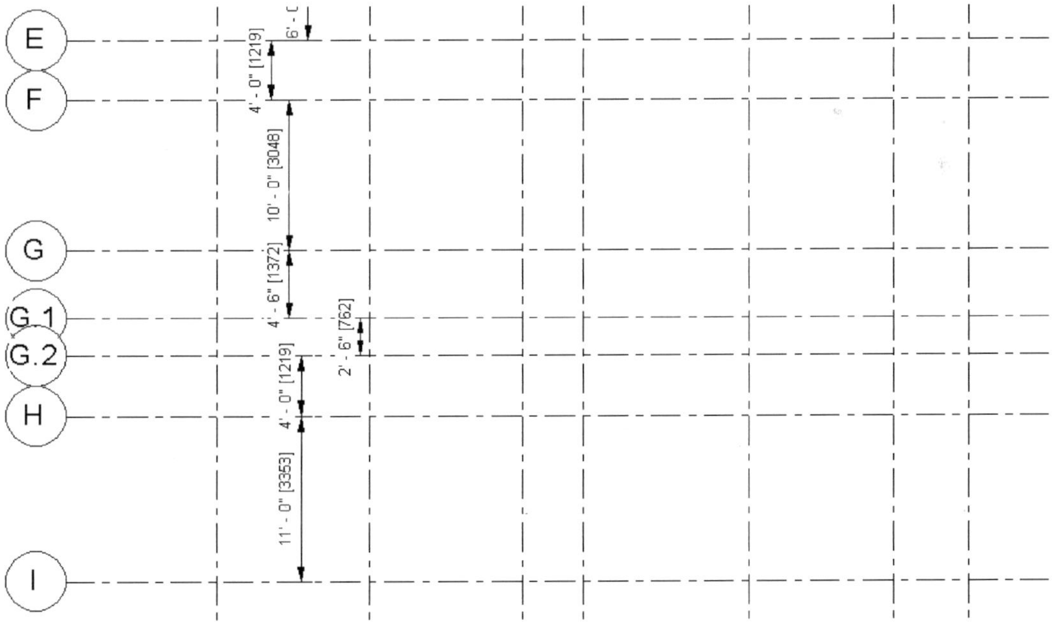

Figure 3-18 *The remaining horizontal grid lines in continuation of Figure 3-17*

Note

*To create the remaining horizontal grids from grid **B** to grid **I** refer to Figures 3-17 and 3-18. Note that both these figures are part of a single grid plan view. Figure 3-18 is in continuation with Figure 3-17 and is displayed separately to show the grid dimensions clearly. Further, the horizontal grids will be renamed using the same method used for vertical grids.*

Saving the Project

In this section, you will save the project file by using the **Save As** tool.

1. To save the project, choose the **Save As > Project** option from the **File** menu. On doing so, the **Save As** dialog box is displayed.

2. In this dialog box, browse to the C:*rst_2020* location and create a folder with name *c03_rst_2020_tut*. Now, open the *c03_rst_2020_tut* folder and enter **c03_Commercial-Complex_tut1** (for Imperial) **or M_c03_Commercial-Complex_tut1** (for Metric) in the **File name** edit box.

3. Choose the **Save** button; the **Save As** dialog box closes and the project file is saved.

Closing the Project

1. To close the project, choose the **Close** option from the **File** menu.

   The file is closed and this completes Tutorial 1 of Chapter 3.

EXERCISE

Exercise 1 Academic Institution

In this exercise, you will create project information for the project created in the Exercise 1 of Chapter 2. Further, you will add levels and grids to the project file. Refer to Figure 3-19 for grid dimensions and names. Use the following project specifications:

(Expected time: 1 hr 15 min)

1. Project file to be used:
   For Imperial *c02_Academic-Institution_exer1.rvt.*
   For Metric *M_ c02_Academic-Institution_exer1.rvt.*

2. Rename Level 1 and Level 2 as **Foundation Level** and **Entry**, respectively.

3. Specify the elevation of **Foundation Level** and **Entry** levels as:
   For Imperial **-5’6” 0”** and **0’ 0”**.
   For Metric **-1676 mm** and **0 mm**, respectively.

4. Story levels to be added:
   For Imperial **Second Floor - Elevation 12’ 6”**
   Roof - Elevation 25’ 0”
   For Metric **Second Floor - Elevation 3810 mm**
   Roof - Elevation 7620 mm

5. Nonstory level to be added:

   For Imperial **T.O.F- Elevation -5'**
   For Metric **T.O.F- Elevation -1524 mm**

6. Project information to be added:

Project Issue Date	**01/01/2020**
Project Status	**Started**
Client Name	**CADSoft Technologies**
Project Address	**525, Andrews Drive Schererville**
Project Name	**Academic Institution**
Project Number	**AI/03/13**

7. Grids to be added to the **Entry** project view using the dimensions and names as shown in Figure 3-19.

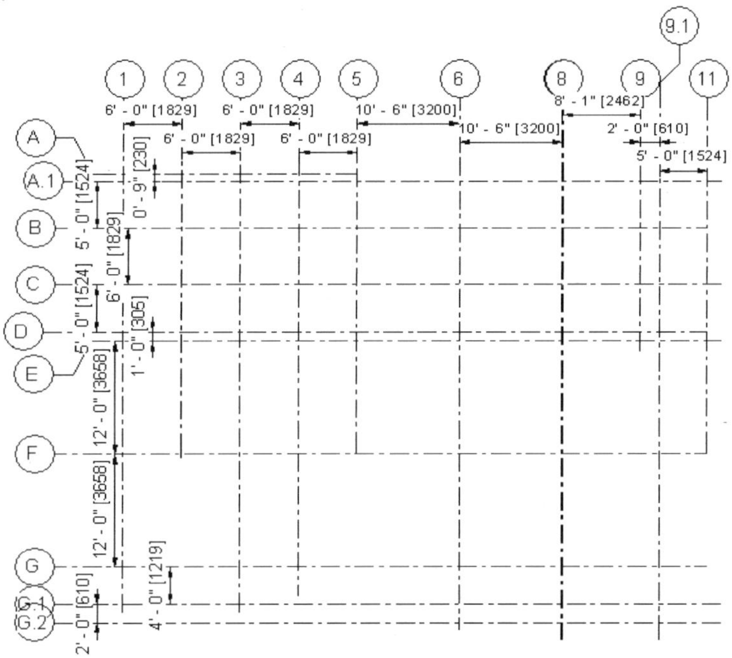

Figure 3-19 *The grid lines with new names for the Academic Institution structure*

8. File name to be assigned:

   For Imperial *c03_Academic-Institution_exer1.rvt.*
   For Metric *M_c03_Academic-Institution_exer1.rvt.*

This page is intentionally left blank

Chapter 4

Structural Columns and Walls

Learning Objectives

After completing this chapter, you will be able to:

- *Create Structural Columns*
- *Modify Structural Columns*
- *Create Structural Walls*
- *Modify Structural Walls*

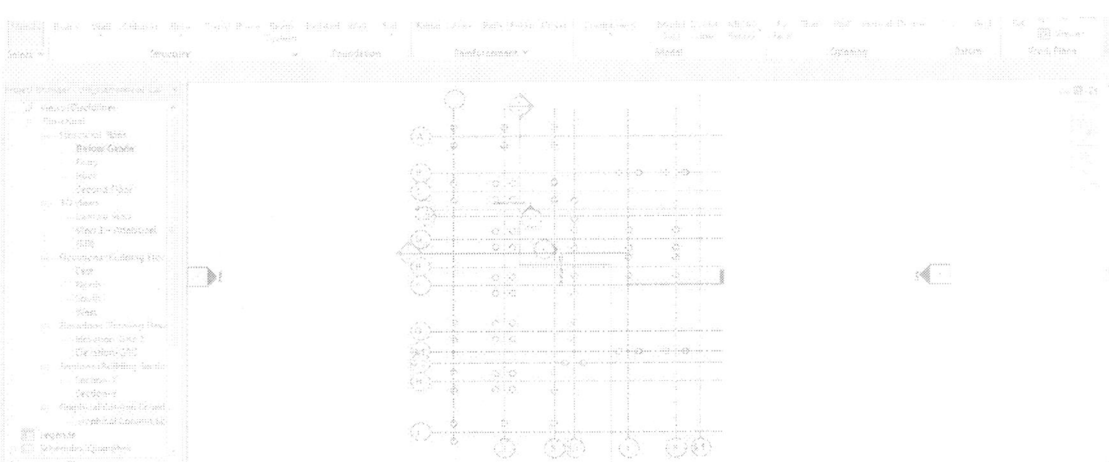

In the previous chapter, you learned to add datums such as structural levels and grids to a project and also learned to add project information to a project. In this chapter, you will learn to add structural elements such as columns and walls. Also, you will learn to work with structural columns and walls in a project in detail.

STRUCTURAL COLUMNS

Structural column is an integral part of the primary structural framing of a building system. The basic function of structural columns is to carry vertical load from the roof system, floor slab, and beam to the foundation. Functionally, columns are subjected to axial loads which form the basis of its design.

In Revit, you need to add columns to a plan view. But before adding columns, it is recommended that you create a grid system. The grid lines and grid intersections of the grid system help you place columns at appropriate locations. When you add columns at grid intersections, they are anchored to it. As a result, whenever there is a change in the grid spacing, the columns move with grid intersections. Before you add structural columns to a project, you need to learn about different types of structural columns. In the following sections, you will learn about different types of structural columns, loading structural columns, adding structural columns and much more.

Types of Structural Column

In Revit, there are three types of structural columns that can be added to a building. These columns are made of three types of material: steel, concrete, and wood. Out of these, steel and concrete columns are commonly used in the construction of a building. While working in Revit, you will get to know that some of the families of these column types are pre-loaded in the project template. You can load other families from the Imperial or Metric structural column libraries that become available on installing Autodesk Revit.

In a project, you can add two types of predefined steel columns depending on your requirement. These columns are rolled steel columns and light gauge steel columns. For rolled steel columns, there are eleven predefined families which are available in the **US Imperial > Structural Columns > Steel** folder location. These families are **Double C-Channel-Column**, **HSS-Hollow Structural Section-Column**, **L-Angle-Column**, **Pipe-Column**, **WWF-Welded Wide Flange-Column**, and more. Similarly, there are other predefined families available in the **US Metric > Structural Columns > Steel** folder location. These families are **CHS-Circular Hollow Section-Column**, **M_L-Angle-Column**, **M_M-Miscellaneous Wide Flange-Column**, **M_W-Wide Flange Column**, **PFC-Parallel Flange Channel-Column**, and **SHS- Square Hollow Section-Column**. For light gauge steel columns, there are seven predefined families which are available at the **US Imperial > Structural Columns > Light Gauge Steel** folder location. The predefined light gauge steel families for columns are **Light Gauge-Channel Studs-Column**, **Light Gauge-Furring Channels-Column**, and **Light Gauge-Studs-Column**. The predefined light gauge steel families for columns available at the **US Metric > Structural Columns > Light Gauge Steel** folder location are **M_Light Gauge-Zeds-Column**.

Note

Apart from using the predefined sections available in Autodesk Revit, you can also customize a section with different physical properties for your project. To use customized sections, first you need to create families and then load them into your project.

There are two types of concrete columns in both US Imperial and US Metric that can be added to a project: casted (in-situ) concrete columns and pre-cast concrete columns. For casted concrete columns in US Imperial, there are several pre-defined families that can be used such as **Concrete-Rectangular-Column with Drop Caps**, **Concrete-Round-Column**, and more. For casted concrete columns in **US Metric**, several pre-defined families available are **M_Concrete-Rectangular-Column with Drop Caps**, **M_Concrete-Rectangular-Column**, and more. For pre-cast concrete columns in **US Imperial**, you can use any of the given predefined families such as **Precast-Rectangular Column with Bearing**, **Precast-Rectangular Column with Multi Layer Corbels**, **Precast-Rectangular Column with Rectangular Fork bearing**, and **Precast-Square Column**. For pre-cast concrete columns in **US Metric**, you can use any of the given predefined families such as **M_Precast-Rectangular Column with Bearing**, **M_Precast-Rectangular Column Cone-Shaped Fork bearing**, and **M_Precast-Square Column**.

To add wood (timber) columns for Imperial project, you can use any of the following families such as **Glulam-Southern Pine-Column**, **Timber-Column**, and more. To add wood (timber) columns in a Metric project, you can use any of the following families: **M_Dimension Lumber-Column**, **M_PSL-Parallel Strand Lumber-Column**, **M_Timber-Column**, and more.

ARCHITECTURAL COLUMNS

Ribbon:	Architecture > Build > Column drop-down > Column: Architectural

Architectural columns are non load bearing members that are placed in a structural project to add ambience to the structure. Generally, very few architectural columns are placed in a structural model. The placing of architectural columns in a model depends upon the design requirement, both architectural and structural.

You can add architectural columns only in plan view. To add an architectural column, activate the desired plan view and then invoke the **Column: Architectural** tool from the **Build** panel; the **Modify | Place Column** contextual tab will be displayed, as shown in Figure 4-1.

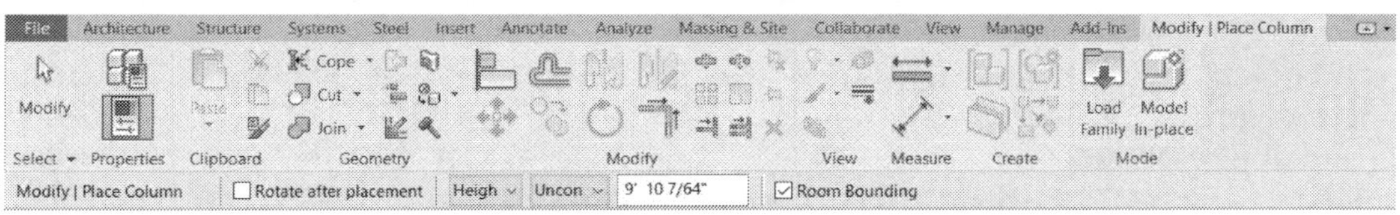

Figure 4-1 *The **Modify / Place Column** contextual tab*

Note

*On invoking the **Column: Architectural** tool, the **Revit** message box may appear informing that no column family is loaded in the project. The message box will prompt you to load a column family in the project. To load a column family, choose the **Yes** button; the **Open** dialog box will be displayed. Browse to the **Column** sub-folder in the **US Imperial** folder for Imperial (for Metric browse to **Column** sub-folder in the **US Metric**) in this dialog box to load the desired column family in your project.*

The tools in the **Modify | Place Column** contextual tab are the same as those discussed in the **Modify | Place Structural Column** contextual tab. After using the tools from the **Modify|**

Place Column contextual tab, you can use options in the **Options Bar** to set the alignment and location of the column that you need to place. After setting the options in the **Options Bar**, click on the desired location in the structural plan view; the architectural column will be placed at the point you click. The placed column can now be used for aesthetic purpose as well as to add structural column to it. Figure 4-2 shows a three-dimensional view of an architectural column placed in a structural model. The method of adding structural column to a architectural column is discussed next.

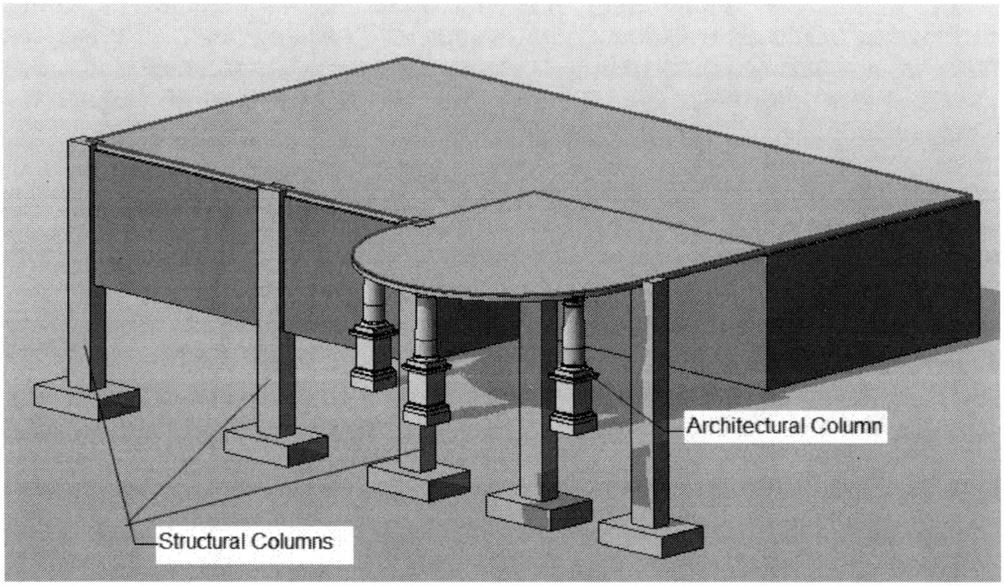

Figure 4-2 *Three-dimensional view of a structural model displaying architectural columns*

Adding Structural Column to an Architectural Column

After adding an architectural column to a project, you can add a structural column to it so that it can become a part of the analytical model. To do so, choose the **Structural Column** tool from **Architecture > Build > Column** drop-down; the **Modify |Place Structural Column** contextual tab will be displayed. In this tab, choose the **At Columns** tool from the **Multiple** panel; the **Modify | Place Structural Column > At Architectural Columns** contextual tab will be displayed. In the **Type Selector** drop-down list in the **Properties** palette, you can select the desired type of column and then click on any existing architectural column from the structural model in the drawing area; the selected structural column type will be placed at the center of the selected architectural column. You can add more structural columns to other architectural columns that are present in the project. To do so, select the architectural columns, press and hold the CTRL key and then click on the architectural columns one by one; the structural columns will be placed at the center of each selected architectural columns. While adding architectural columns to the selection, you can also remove the elements from the selection. To do so, press and hold the SHIFT key and click on the selected column that you need to remove; the selected structural column will be removed from the selected columns.

After placing the structural column(s) in the architectural column(s), choose the **Finish** button from the **Multiple** panel; the structural column(s) will be placed at the desired architectural columns.

STRUCTURAL WALLS

A wall constitutes an important feature in a structural model. Walls can be modeled as a partition, boundary element, a shear resistant element, or a load bearing element. Walls differ from each other on the basis of their structural usage. The structural usage of a wall defines its load-bearing functions used for structural and architectural design.

A building project contains different types of walls depending on their structural usage. These walls can be exterior walls, interior walls, basement walls, and partition wall studs. In Revit, walls are defined as system families and therefore, they can be created and modified within the current structural model. The walls are implicit features in a structural model. This means that they cannot be added in the model as separate entities and also they change with the change in the structural model.

In Revit, you can add structural walls that are load-bearing in function. This implies that they play an active role in transferring the load from one structural element to the other. They can bear both vertical forces and shear forces depending upon the function they are supposed to perform. The function of a structural wall is determined by its structural usage that can be defined in its instance property.

TUTORIAL

Tutorial 1 Commercial Complex

In this tutorial, you will add structural columns and structural walls to the *Commercial Complex* project created in Tutorial 1 of Chapter 3. Use the following project parameters and specifications for completing the project. **(Expected time: 1 hr)**

1. Structural column:
   - For Imperial: Type: **Concrete-Rectangular-Column**; Dimension: 12" x 24"
   - For Metric: Type: **M_Concrete-Rectangular-Column**; Dimension: 300 x 600mm

2. Structural concrete walls:
   - For Imperial: Width - **12"**; Function - **Core-shaft**
   - For Metric: Width - **300mm**; Function - **Core-shaft**

3. File name to be assigned:
   - For Imperial: *c04_Commercial-Complex_tut1.rvt.*
   - For Metric: *M_c04_Commercial-Complex_tut1.rvt.*

The following steps are required to complete this tutorial:

a. Start a Revit session and open the project file.
   For Imperial *c03_Commercial-Complex_tut1.rvt*
   For Metric *M_c03_Commercial-Complex_tut1.rvt*
b. Create a structural column type.
c. Add columns.
d. Modify the height of columns.
e. Add structural walls.
f. Save the project using the **Save As** tool.
g. Close the project using the **Close** tool.

Opening the Project

In this section, you will open the project file created in Tutorial 1 of Chapter 3.

1. Choose **Open > Project** from the **File** menu; the **Open** dialog box is displayed.

2. In this dialog box, browse to the *C:\rst_2020\c03_rst_2020_tut* location and then choose the *c03_Commercial-Complex_tut1.rvt* (for Imperial) *or M_c03_Commercial-Complex_tut1.rvt* (for Metric) project file. You can also download this file from *https://www.cadsofttech.com*. The path of the file is as follows: *Textbooks > Civil/GIS > Revit Structure > Revit Structure 2020 for Novices*.

3. Choose the **Open** button from the **Open** dialog box; the selected project file opens in the drawing window in the **Entry** structural plan view.

Creating a Structural Column Type

In this section, you will create a concrete rectangular column type to add to the project view.

1. Choose the **Column** tool from the **Structure** panel of the **Structure** tab; the **Modify | Place Structural Column** contextual tab is displayed.

2. In the **Properties** palette, select the specified option from the **Type Selector** drop-down list.
   For Imperial **Concrete-Rectangular-Column: 12 x 18**
   For Metric **M_Concrete-Rectangular-Column: 300 x 450mm**

3. Choose the **Edit Type** button; the **Type Properties** dialog box is displayed.

4. In this dialog box, choose the **Duplicate** button; the **Name** dialog box is displayed.

5. In the **Name** dialog box, enter **12 x 24** (in Imperial unit system) and **300 x 600mm** (in Metric unit system) in the **Name** edit box and choose **OK**; the **Name** dialog box is closed.

6. In the **Type Properties** dialog box, click in the **Value** field corresponding to the **h** parameter; the existing value, **1' 6"** (for Imperial) and **450.0** (for Metric), is highlighted.

7. Replace the existing value with **2' 0"** (for Imperial) and **600.0** (for Metric), and then choose the **OK** button; the **Type Properties** dialog box closes and preview of the new column type attached with the cursor is displayed in the drawing area.

Adding Structural Columns

In this section, you will add structural columns to the model.

1. In the drawing area, right-click and then choose the **Zoom In Region** option from the shortcut menu; the cursor changes to a zoom icon.

2. Move the cursor and place it between the grid lines marked **1** and **A**. Click at an appropriate location, as shown in Figure 4-3; a rectangular selection window appears.

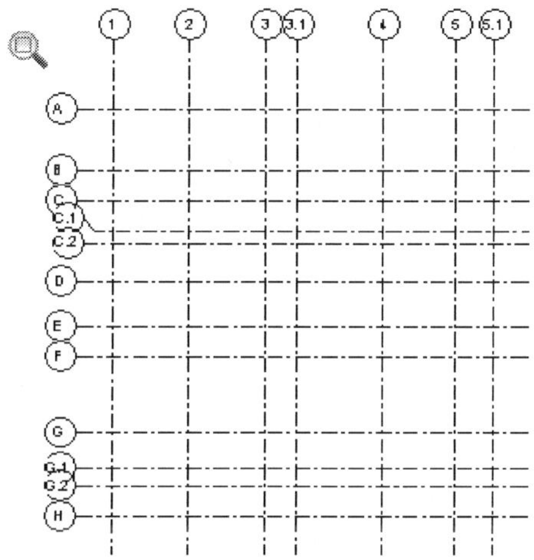

Figure 4-3 *Placing the cursor to mark the first corner of the zoom window*

3. Move the cursor downward toward right. Click between the grid lines marked 5.1 and C.2, refer to Figure 4-4. On doing so, the project view zooms in so that the columns can easily be added at the desired location.

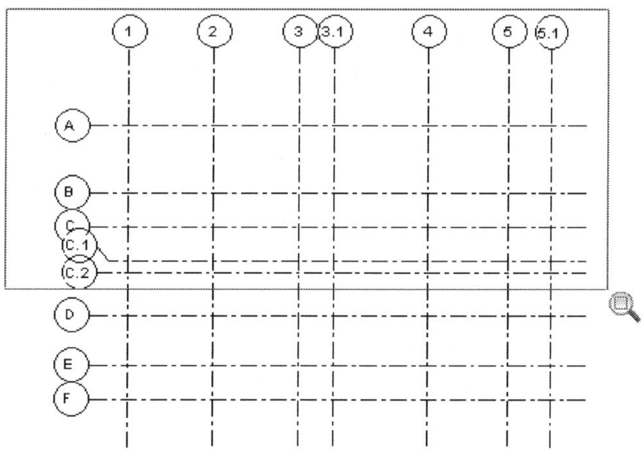

Figure 4-4 *Placing the cursor to mark the other corner of the zoom window*

4. In the **Options Bar**, select the **Rotate after placement** check box and then select the **Height** option from the first drop-down list next to it.

5. Ensure that the **Room Bounding** check box is selected in the **Options Bar**. Place the cursor at the intersection of grid 1 and grid A; the grid 1 and grid A are highlighted.

6. Click when both grids are still highlighted; a preview of the column with its center aligned to the intersection of grid 1 and grid A is displayed.

7. Move the cursor up toward the right; a rotation alignment line appears with temporary angular dimension.

8. Type **90**; an edit box with typed value is displayed near the temporary dimension.

9. Press ENTER; a column is inserted at the intersection of grid 1 and grid A, as shown in Figure 4-5.

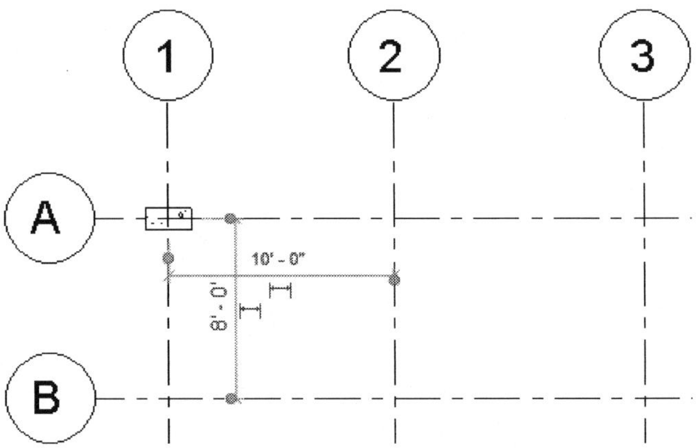

***Figure 4-5** A column added at the intersection of grid 1 and grid A*

10. Repeat steps 5 to 9 to insert other columns of the same type and alignment at different grid intersections, refer to Figure 4-6. Do not exit the **Column** tool even though you have inserted all columns at different grid intersections.

Note

While inserting columns at different grid intersections, you can zoom out the current view and then zoom into the area where you need to insert columns.

11. After inserting all the columns, choose the **At Grids** tool from the **Multiple** panel of the **Modify | Place Structural Column** contextual tab; the **Modify | Place Structural Column > At Grid Intersection** contextual tab is displayed.

12. Place the cursor on grid B. Click when the grid is highlighted; the grid B is selected.

13. Press and hold the CTRL key and click on grid 4, grid 5, and grid G.1 to select them; columns are inserted at the grid intersections of the selected grids, as shown in Figure 4-7.

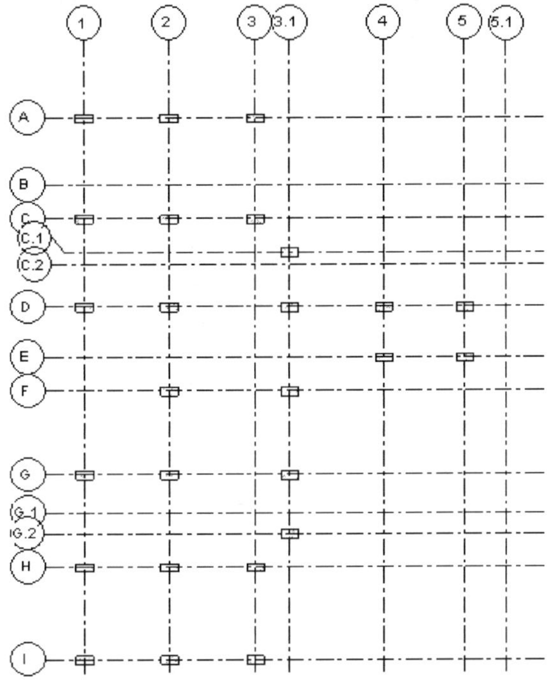

Figure 4-6 *Alignment and columns of same type inserted in the grid*

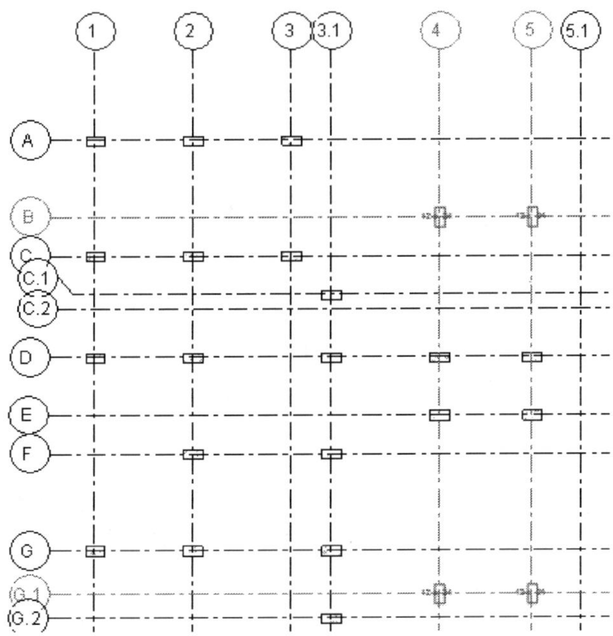

Figure 4-7 *Columns added at the intersections of the highlighted grids*

14. Now, choose the **Finish** button from the **Multiple** panel in the **Modify | Place Structural Column > At Grid Intersection** contextual tab; the **Modify | Place Structural Column** contextual tab is displayed.

15. Choose the **Modify** button from the **Select** panel in the **Modify | Place Structural Column** contextual tab.

Modifying the Height of Columns

In this section, to adjust the thickness of the slab (floor) in the structural model, you will modify the height of the inserted columns (except for stairs and lifts) by adjusting their top offset distances.

1. Place the cursor between grid A and grid 1, refer to Figure 4-8.

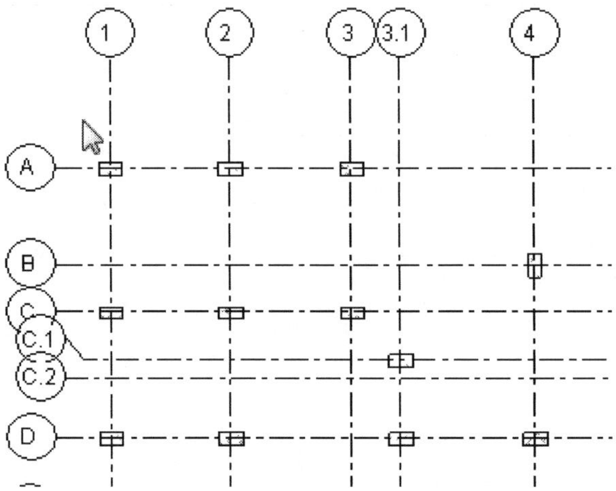

Figure 4-8 *Specifying the first point of the selection window*

2. Next, press and hold the left mouse button and then drag the cursor along the selection window downward toward the right until it is placed at the location shown in Figure 4-9.

3. Release the mouse button and click; all the columns inside the selection window are selected and the **Modify | Structural Columns** contextual tab is displayed.

4. In the **Properties** palette of the selected columns, click in the value field corresponding to the **Top Offset** instance parameter. Clear the existing value and then enter **-1'** (in Imperial) and **-304.8mm** (in Metric).

5. Next, choose the **Apply** button in the **Properties** palette and then the **Modify** button in the **Select** panel of the **Modify | Structural Columns** contextual tab.

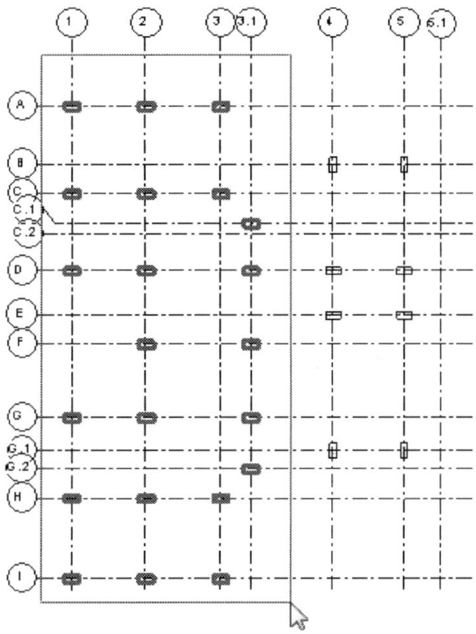

Figure 4-9 *Specifying the other point of the selection window*

6. Now, select all the columns using the selection window, as done earlier.

7. In the **Properties** palette of the selected columns, select the **Entry** option for the **Base Level** parameter and the **Second Floor** option for the **Top Level** parameter if not selected.

8. Choose the **Apply** button; the height of the columns is set between the **Entry** and **Second** floor levels.

Adding Structural Walls

In this section, you will create a structural wall type and add it to the project view.

1. Choose the **Wall: Structural** tool from **Structure > Structure > Wall** drop-down; the **Modify | Place Structural Wall** contextual tab is displayed.
2. In the **Properties** palette, select the desired option from the **Type Selector** drop-down list.
   For Imperial **Exterior - 10" Concrete**
   For Metric **Exterior - 225mm Concrete**
   Now, choose the **Edit Type** button; the **Type Properties** dialog box is displayed
3. Choose the **Duplicate** button in this dialog box; the **Name** dialog box is displayed.
4. Type the specified values in the **Name** edit box of the **Name** dialog box.
   For Imperial **Core-Shaft - 12" Concrete**
   For Metric **Core-Shaft - 300mm Concrete**
   Next, choose **OK**; the **Name** dialog box is closed.

5. In the **Type Properties** dialog box, choose the **Edit** button displayed in the **Value** column of the **Structure** parameter; the **Edit Assembly** dialog box is displayed.

6. In the **Layers** area of the **Edit Assembly** dialog box, click in the edit box in the **Thickness** column corresponding to the **Structure [1]** function and enter **12**"(for Imperial) and **300mm** (for Metric). Next, press ENTER to replace the existing value.

7. Choose the **OK** button; the **Edit Assembly** dialog box closes and the **Type Properties** dialog box is displayed.

8. In the **Type Properties** dialog box, select the **Core-shaft** option from the drop-down list corresponding to the **Function** parameter from the **Construction** area and then choose the **OK** button; the **Type Properties** dialog box is closed.

9. In the **Options Bar**, select the **Height** option from the first drop-down list; a warning message box appears at the bottom of the screen. Close the message box.

10. Select the **Second Floor** option from the second drop-down list in the **Options Bar** and then clear the **Chain** check box.

11. Zoom in the drawing and move the cursor toward the column at the intersection of grid 4 and grid B. Place the cursor on the right face of the column. Click when the right face is highlighted and the **Midpoint** snap is displayed, as shown in Figure 4-10.

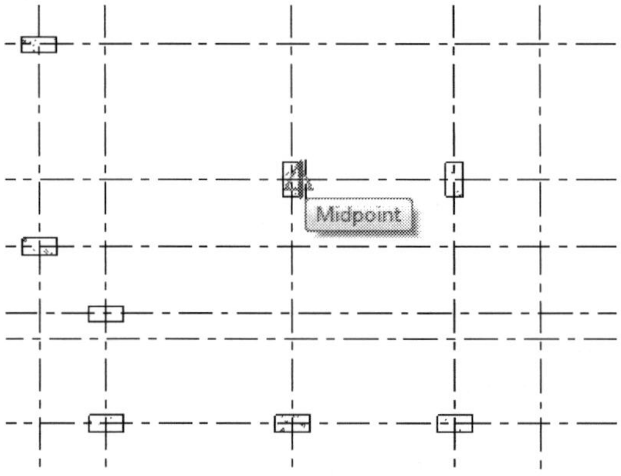

Figure 4-10 *Specifying the start point at the intersection of grid 4 and grid B*

12. Now, move the cursor horizontally toward the right and place it at the midpoint of the left face of the column located at the intersection of grid 5 and grid B, as shown in Figure 4-11. Click when the **Midpoint and Horizontal** snap appears.

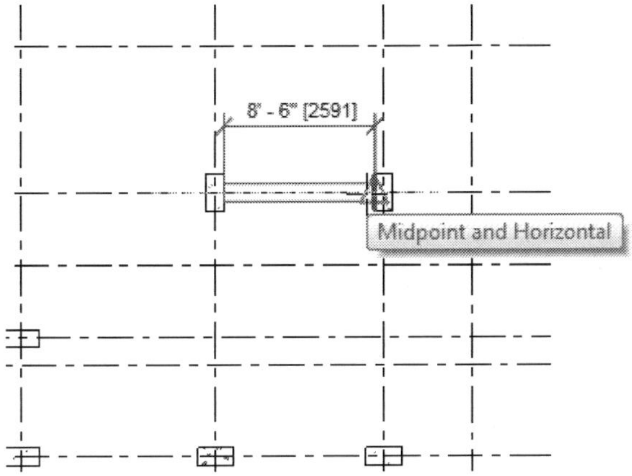

Figure 4-11 *Specifying the endpoint at the intersection of grid 5 and grid B*

13. Move the cursor and place it at the lower face of the column, located at the intersection of grid 5 and grid B. Click when the **Midpoint** snap appears, as shown in Figure 4-12.

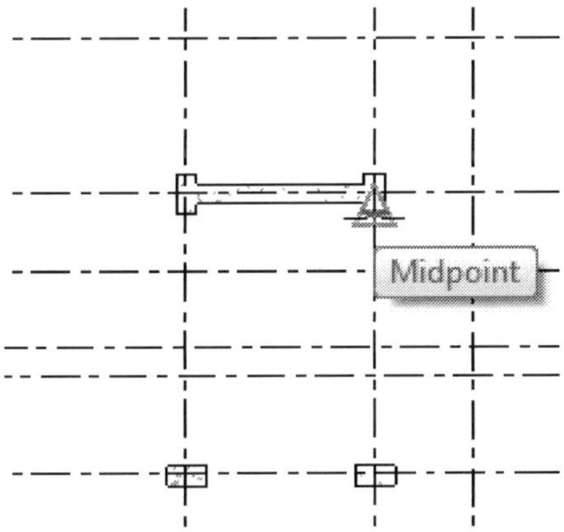

Figure 4-12 *Specifying the start point of the other wall at the midpoint of the lower face of the column at the intersection of grid 5 and grid B*

14. Move the cursor vertically down and place it at the upper face of the column at the intersection of grid 5 and grid D. Click when the **Endpoint and Vertical** snap appears.

15. Next, place the cursor at the lower face of the column, at the intersection of grid 5 and grid D. Click when the lower face is highlighted and the **Midpoint** snap appears.

16. Move the cursor vertically downward and place it at the upper face of the column located at the intersection of grid 5 and grid E. Click when the **Endpoint and Vertical** snap appears at the intersection point.

17. Repeat the method of adding structural walls discussed in Steps 13 to 16 to add the remaining structural walls to the project, as shown in Figure 4-13.

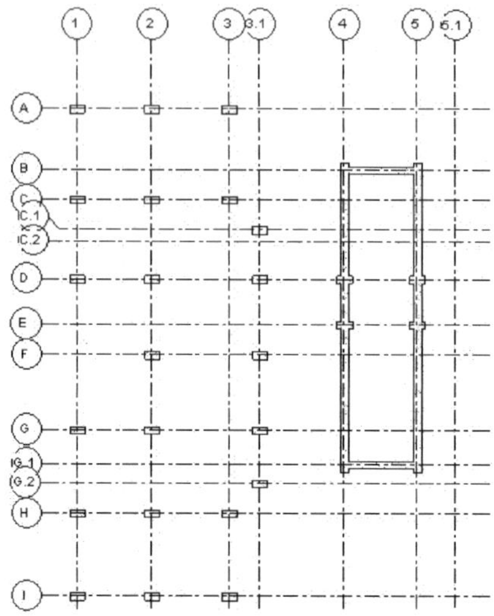

Figure 4-13 *The structural walls added*

18. Choose the **Modify** button from the **Select** panel to exit the **Wall: Structural** tool.

19. Select the wall placed on grid Band then right click; a shortcut menu is displayed. Choose the **Select All Instances > Visible in View** from the shortcut menu; all walls in the project view will be highlighted.

20. In the Properties palette, enter **-2'** in the **Top Offset** edit box.

Saving the Project

In this section of the tutorial, you will save the project file using the **Save As** tool.

1. Choose **Save As > Project** from the **File** menu. On doing so, the **Save As** dialog box is displayed.

2. In this dialog box, browse to the *C:\rst_2020* location and create a folder with the name *c04_rst_2020_tut*. Next, open the *c04_rst_2020_tut* folder and enter **c04_Commercial-Complex_tut1** (for Imperial) and **M_c04_Commercial-Complex_tut1** (for Metric) in the **File name** edit box.

3. Choose the **Save** button; the **Save As** dialog box closes and the project file is saved.

Closing the Project

1. Choose the **Close** option from the **File** menu.

EXERCISE

Exercise 1 Academic Institution

In this exercise, you will add structural columns with tags and structural walls to the exercise file created in Exercise 1 of Chapter 3, refer to Figure 4-14. Use the following project parameters and specifications: **(Expected time: 1 hr 15 min)**

1. Structural column:
   For Imperial Type: **Concrete-Rectangular-Column**
   Dimension: **18" x 24"**
   For Metric Type: **M_Concrete-Rectangular-Column**
   Dimension: **450 x 600mm**
2. Structural wall:
   For Imperial Type: **Exterior-10" Concrete**.
   For Metric Type: **Exterior-250mm Concrete**.
3. Project file to be used:
   For Imperial *c03_Academic-Institution_exer1.rvt.*
   For Metric *M_c03_Academic-Institution_exer1.rvt*
4. Base and Top constraints for walls and columns are **Entry** and **Second Floor** levels.
5. Columns and walls are to be added in the **Entry** structural plan view.
6. File name to be assigned:
   For Imperial *c04_Academic-Institution_exer1.rvt.*
   For Metric *M_c04_Academic-Institution_exer1.rvt*
7. For locations and placement of columns and walls, refer to Figure 4-53.

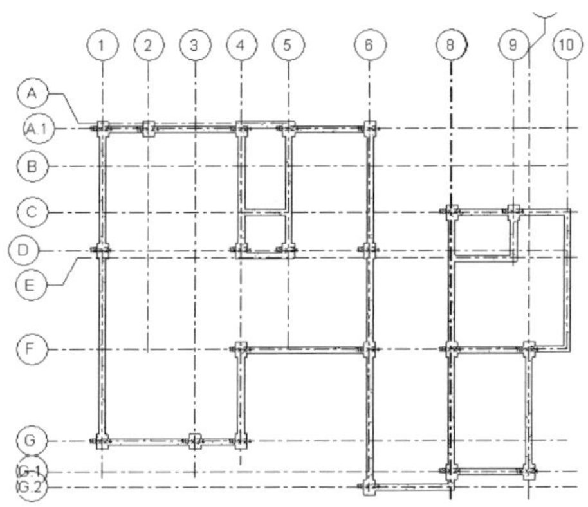

Figure 4-14 *The walls and columns added to the Academic-Institution project*

This page is intentionally left blank

Chapter 5

Foundations, Beams, Floors, and Open Web Joists

Learning Objectives

After completing this chapter, you will be able to:
- *Create foundations*
- *Add structural floors*
- *Add beams*

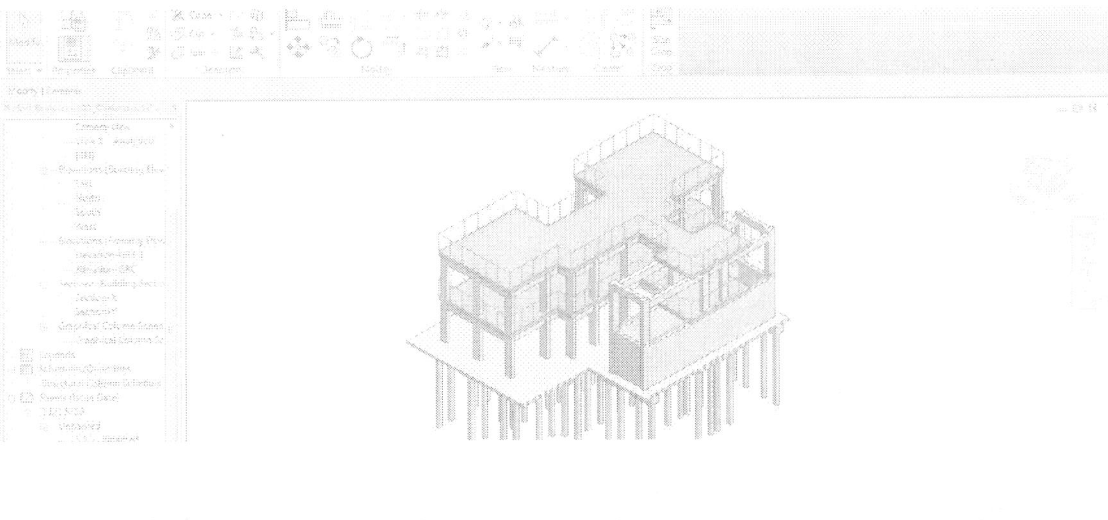

In the previous chapter, you learned to add different structural components such as walls, columns, and so on. In this chapter, you will learn about different types of foundations. You will also understand how to add and modify beams and structural floor and later on add them in your project. Moreover, you will learn to add slab edge to floor and also about the open web steel joists.

FOUNDATIONS

Foundation is one of the important components in a structural model. The function of foundation in any of the construction project is to provide support to the entire building. Foundation acts as an interface between building structure and earth strata for transferring load. For a structural model, you can either have a shallow or deep foundation depending on the bearing capacity of the soil on which the building structure is to be erected. Shallow foundation can be further subcategorized as strip footing and raft footing while deep foundation as pile foundation. Depending on soil bearing capacity (SBC) and structure we can provide either of them.

In Autodesk Revit , you can add foundations of different types such as Isolated, Pile, Wall, and Slab. These types represent different physical foundations that are applicable to sites such as isolated footing (for columns), retaining wall, stepped footing, bearing footings and raft foundation. In this chapter, you will learn how to add both the shallow and deep foundations.

Foundation Wall

After adding structural walls, you can add a foundation wall to the existing structural model. This wall serves as a basement wall or a retaining wall that goes down below the ground level to a level from where the top of the foundation starts. To add a foundation wall, first you need to create the required levels to constrain the ends of the wall. For example, if you want to create a basement wall for an existing structural model, you need to have two levels, one level to constrain the top of the basement wall and the other level to constrain the bottom of the basement wall. The top of the basement wall may be at a level above or at ground level whereas the bottom of this wall will be at a level that coincides with the top of the footing. Figure 5-1(a) and Figure 5-1(b) display a basement wall with its top constrained to the **Entry Level** level and its bottom constrained to the **T.O. Footing** level in the imperial and metric forms respectively.

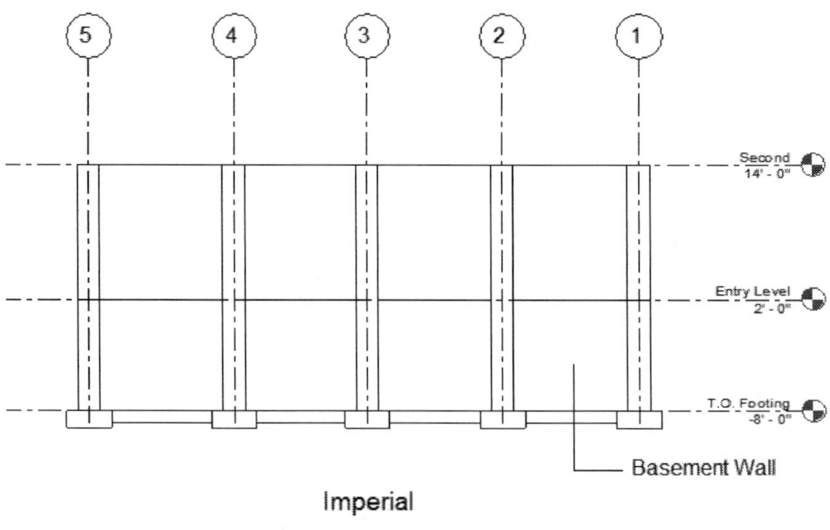

Figure 5-1(a) *The basement wall with the top and bottom constraints (Imperial)*

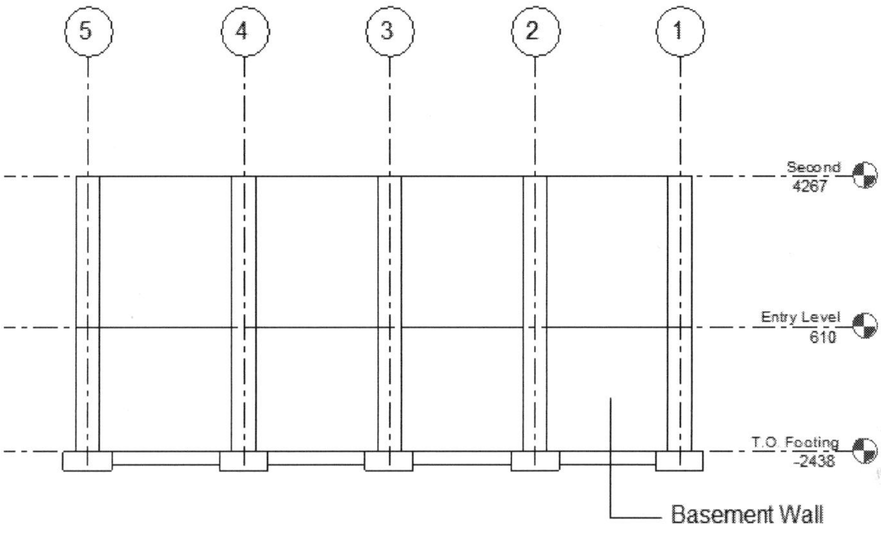

Figure 5-1(b) The basement wall with the top and bottom constraints (Metric)

Shallow Foundations

Shallow foundations are used to transfer load from a building to the earth strata near the ground surface. Shallow foundations include strip footing foundations, spread footing foundations, mat-slab, and slab-on-grade foundations. In Revit , you can add a shallow foundation as a wall foundation (strip footing foundation), isolated foundation (spread footing foundation) for columns, or foundation slabs. To add these foundations, you need to invoke various tools such as **Isolated**, **Wall**, and **Foundation Slab** from the **Foundation** panel in the **Structure** tab. Figure 5-2 shows various tools that can be invoked to add different types of foundations to a structural model. In the **Foundation** panel, you can invoke the **Wall** tool to add a wall foundation in the existing structural model. You can use this tool to add two types of wall foundations: Bearing footing foundation and Retaining footing foundation. Each of these foundations has separate element properties and placement methods. These properties and methods will be discussed later in this chapter. You can use the **Isolated** tool to add the spread or isolated foundations to the columns of the structural model. The element properties and the placement method for this type of foundation will also be discussed later. Note that the wall and the isolated footings are dependent on the host element to which they are attached. Figure 5-3 shows the associativity of a structural wall with a wall foundation.

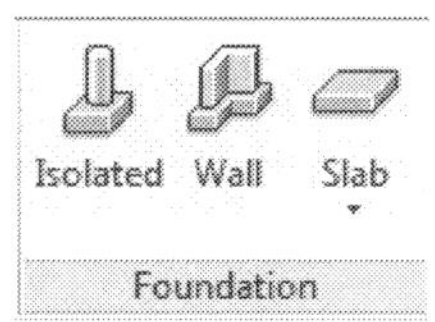

***Figure 5-2** Various tools to add foundation*

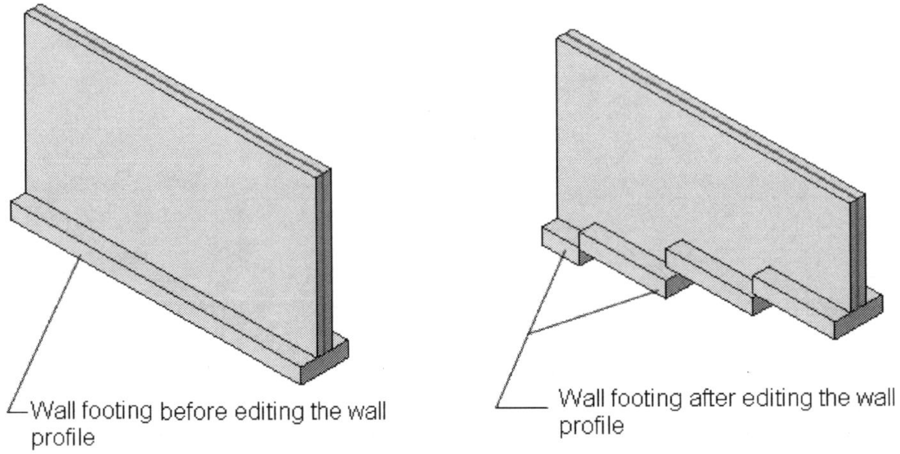

Figure 5-3 *The wall foundation before and after editing the wall profile*

The **Slab** tool in the **Foundation** panel can be used to add a foundation slab to the structural model. Unlike the isolated and wall foundations, the slab foundations are not attached to any host element. Slab foundations are created independent of other structural elements present in a structural model. The placement and the properties of this type of foundation will be discussed later in this chapter.

STRUCTURAL FLOORS

The structural floor in a structural model represents a floor deck or a floor slab. In Autodesk Revit, you can specify various structural parameters such as thickness, structural composition, rebar sizes, and more to the added structural floor. You can also add slope and create opening in the added structural floor. Figure 5-4 shows a structural floor in a model. In the next sections, you will learn how to add structural floors.

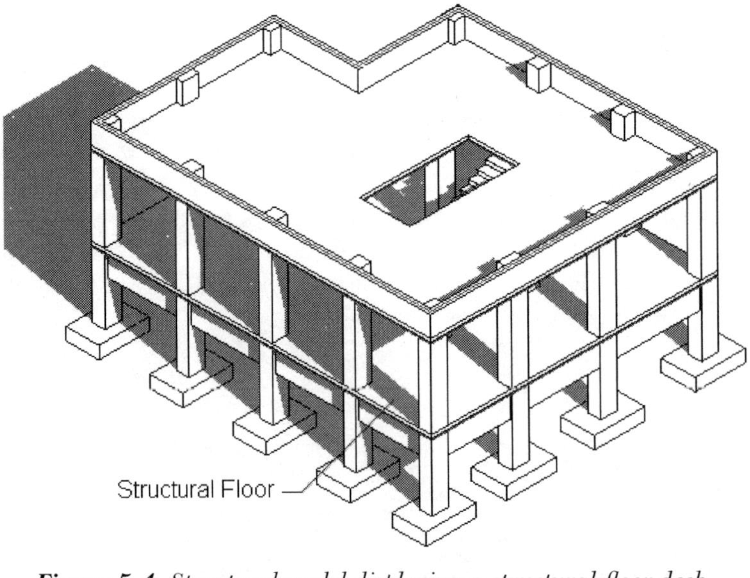

Figure 5-4 *Structural model displaying a structural floor deck*

BEAMS

Beams are structural elements that can be used to resist load. Beams generally carry vertical forces. They can also carry horizontal forces as that of wind and earthquake. Figure 5-5 shows steel beams between vertical steel columns.

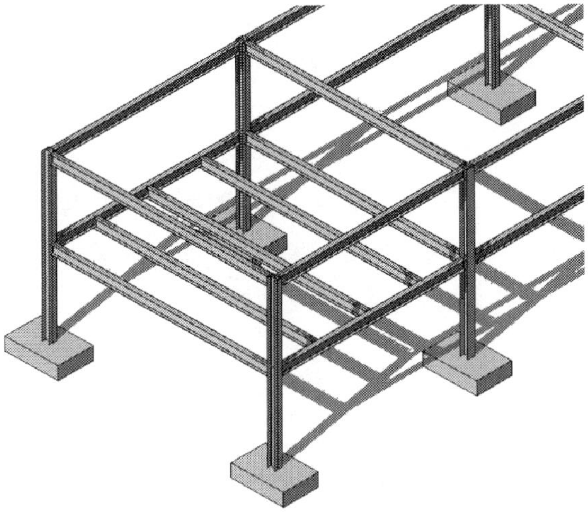

Figure 5-5 *Steel beams between vertical steel columns*

In Autodesk Revit , you can add different families of beams to your structural model. You can assign different properties to a beam by assigning various instance and type parameters to it. In the following sections, you will learn how to add and modify beams.

TUTORIAL

Tutorial 1 Commercial Complex

In this tutorial, you will add foundations, foundation slab, and structural framing members to the *Commercial Complex* project created in Tutorial 1 of Chapter 4.

(Expected time: 1 hr 45 min)

Use the following project parameters and specifications:

1. Foundations:

For Imperial Types: **Pile Cap-2 : Pile 31"x61"x35"**
Pile Cap-4 : Pile 72"x72"x35"
Wall Foundation : Bearing Footing -36"x12"

For Metric Types: **M_Pile Cap-2 : Pile 800 x 1500 x 900mm**
M_Pile Cap-4 : Pile 1800 x 1800 x 900mm
Wall Foundation : Bearing Footing - 900 x 300mm

2. Foundation Slab:

For Imperial Type: **Foundation Slab : 12" Foundation Slab**
For Metric Type: **Foundation Slab : 300mm Foundation Slab**

3. Beams:

For Imperial	Type: **Concrete-Rectangular Beam : 12 x 24**
For Metric	Type: **M_Concrete-Rectangular Beam: 300 x 600mm**

4. Structural Floor:

For Imperial	Type: **Generic 12"**
For Metric	Type: **Generic 300mm**

5. File name to be assigned:

For Imperial	*c05_Commercial-Complex_tut1.rvt.*
For Metric	*M_c05_Commercial-Complex_tut1.rvt.*

The following steps are required to complete this tutorial:

a. Open the *c04_Commercial-Complex_tut1.rvt* project file.
b. Load the isolated foundation families.
c. Create new types for isolated foundations.
d. Add isolated foundation.
e. Add foundation wall.
f. Add foundation slab.
g. Add beams.
h. Modify beams.
i. Add structural floor.
j. Save the project using the **Save As** tool.
k. Close the project using the **Close** tool.

Opening the Existing Project

In this section, you will open the project file created in Tutorial 1 of Chapter 4.

1. To open the project file, choose **Open > Project** from the **File** menu; the **Open** dialog box is displayed.

2. In this dialog box, browse to the *C:\rst_2020\c04_rst_2020_tut* location and choose the project file *c04_Commercial-Complex_tut1.rvt* for Imperial unit system and *M_c04_Commercial-Complex_tut1.rvt* for Metric unit system. You can also download this file from *https://www.cadsofttech.com*. The path of the file is as follows: *Textbooks > Civil/GIS > Revit Structure > Revit Structure 2020 for Novices*.

3. Choose the **Open** button from the **Open** dialog box; the selected project file opens in the drawing window in the **Entry** structural plan view.

Loading Isolated Foundation Families

In this section, you will load the isolated foundation families to the current project.

1. Ensure that the **Entry** structural plan view is the current view in the drawing area. Choose the **Isolated** tool from the **Foundation** panel of the **Structure** tab; the **Modify | Place Isolated Foundation** contextual tab is displayed.

2. In this tab, choose the **Load Family** tool from the **Mode** panel; the **Load Family** dialog box is displayed.

3. In the **Load Family** dialog box, browse to folder location.
   For Imperial **US Imperial > Structural Foundations**
   For Metric **US Metric > Structural Foundations**

4. Press and hold the CTRL key and then click on the *Pile Cap-2 Pile.rfa* and *Pile Cap-4 Pile.rfa* files displayed in the **Structural Foundations** folder for Imperial or select *M_Pile Cap-2 Pile.rfa* and *M_Pile Cap-4 Pile* for Metric system; the files are highlighted indicating they have been selected.

5. Choose the **Open** button in the **Load Family** dialog box; the window closes and the highlighted family files are loaded in the project.

Creating New Types for the Isolated Foundation

In this section, you will create two new types of the isolated foundation.

1. In the **Properties** palette, ensure that the **Pile Cap-2 : Pile 31"x71"x35"** option for Imperial unit system or **M_Pile Cap-2 : Pile 800 x 1800 x 900mm** option for Metric unit system is selected from the **Type Selector** drop-down list.

2. Choose the **Edit Type** button from the **Properties** palette; the **Type Properties** dialog box is displayed.

3. Choose the **Duplicate** button in the **Type Properties** dialog box; the **Name** dialog box is displayed.

4. In the **Name** dialog box, enter the specified name in the **Name** edit box-
   For Imperial **31"x61"x35"**
   For Metric **800 x1500 x 900**
   Next, choose the **OK** button; the **Name** dialog box is closed and the **Type Properties** dialog box is displayed.

5. In the **Type Properties** dialog box, select the desired option, if not selected by default, from the drop-down list in the **Value** field of the **Pile Type<Structural Foundations>** parameter.
   For Imperial **Pile-Steel Pipe : 16" Diameter**
   For Metric **M_Pile-Steel Pipe : 400mm Diameter**

6. Click in the **Value** field of the **Width** parameter and replace the existing value with **5' 1"** for Imperial unit system or **1500mm** for Metric unit system and press ENTER.

7. Click in the **Value** field of the **Clearance** parameter and replace the existing value with **1'** for Imperial or **300mm** for Metric and then press ENTER. Next, choose the **OK** button; the family is created.

8. Choose the **Edit Type** button from the **Properties** palette; the **Type Properties** dialog box is displayed. Select the **Pile Cap-4 Pile** option from the **Family** drop-down list and then select the desired option from the **Type** drop-down list, if not selected by default.
    For Imperial **79"x 79" x35"**
    For Metric **2000 x 2000 x 900mm**

9. Choose the **Duplicate** button; the **Name** dialog box is displayed.

10. In the **Name** edit box of the **Name** dialog box, enter the specified value-
    For Imperial **72"x 72" x35"**
    For Metric **1800 x1800 x900mm**

    Next, choose the **OK** button, the **Name** dialog box is closed and the **Type Properties** dialog box is displayed.

11. In the **Type Properties** dialog box, select the desired option, if not selected by default, from the drop-down list in the **Value** field of the **Pile Type<Structural Foundations>** parameter.
    For Imperial **Pile-Steel Pipe : 16" Diameter**
    For Metric **M_Pile-Steel Pipe : 400mm Diameter**

12. Click in the **Value** field corresponding to the **Width** and **Length** parameters and enter **6'** in Imperial or **1800mm** in Metric unit systems in each of them.

13. Click in the **Value** field corresponding to the **Clearance** parameter and enter **1' 3"** for Imperial or enter **380mm** for Metric to replace the existing value.

14. Choose the **OK** button; the **Type Properties** dialog box is closed.

Adding Isolated Foundations

In this section, you will add the isolated foundation types that you have created in the project.

1. In the **Properties** palette, select the option from the **Type Selector** drop-down list.
    For Imperial **Pile Cap-4 Pile: 72"x72"x35"**
    For Metric **M_Pile Cap-4 Pile: 1800 x1800 x900mm**

2. Choose the **At Columns** tool from the **Multiple** panel of the **Modify | Place Isolated Foundation** contextual tab.

3. Move the cursor and place it over the column located at the intersection of grid C and grid 2.

4. Click when the column is highlighted; a foundation with four piles and a pile cap is added at the center of the highlighted column.

5. Press and hold the CTRL key and repeat steps 3 and 4 to add foundation to all the internal columns. Refer to Figure 5-6 for the location of foundations.

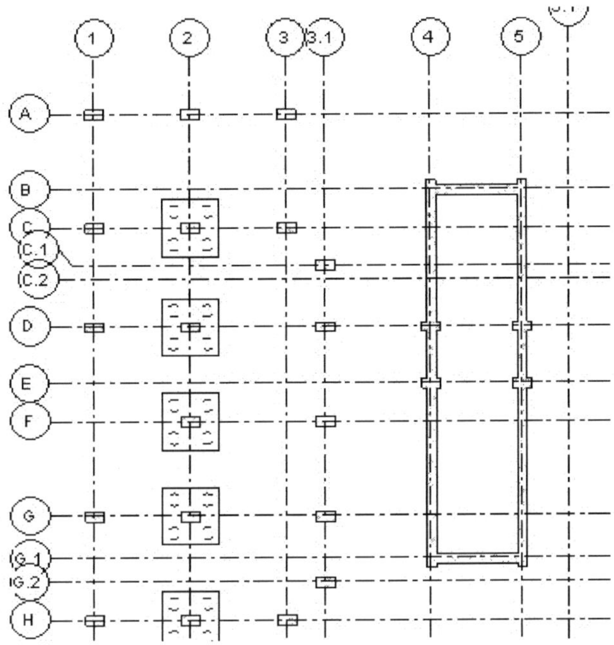

Figure 5-6 *The 4-pile foundations added to all internal columns*

6. Release the CTRL key and choose the **Finish** button from the **Multiple** panel of the **Modify |Place Isolated Foundations > At Structural Column** contextual tab; the 4 pile foundation is added to the selected columns. Ignore if any warning message is displayed. Next, press ESC.

7. Choose the **Isolated** tool from the **Foundation** panel of the **Structure** tab to add the foundations to the remaining columns.

8. In the **Properties** palette, select the option from the **Type Selector** drop-down list.
   For Imperial **Pile Cap -2 Pile : 31"x61"x35"**
   For Metric **M_Pile Cap -2 Pile : 800 x1500 x900mm**

9. Choose the **At Columns** tool from the **Multiple** panel of the **Modify | Place Isolated Foundation** contextual tab.

10. Next, move the cursor and place it over the column located at the intersection of grid 1 and grid A; the column at the intersection of the grids is highlighted.

11. Click on the highlighted column; the foundation is added to the highlighted column.

12. Next press and hold the CTRL key and repeat steps 10 and 11 to add the foundation type to all the columns that do not have a foundation attached to them (except the columns at the intersection of grid C.1 and grid 3.1 and at the intersection of grid G.2 and grid 3.1), refer to Figure 5-7.

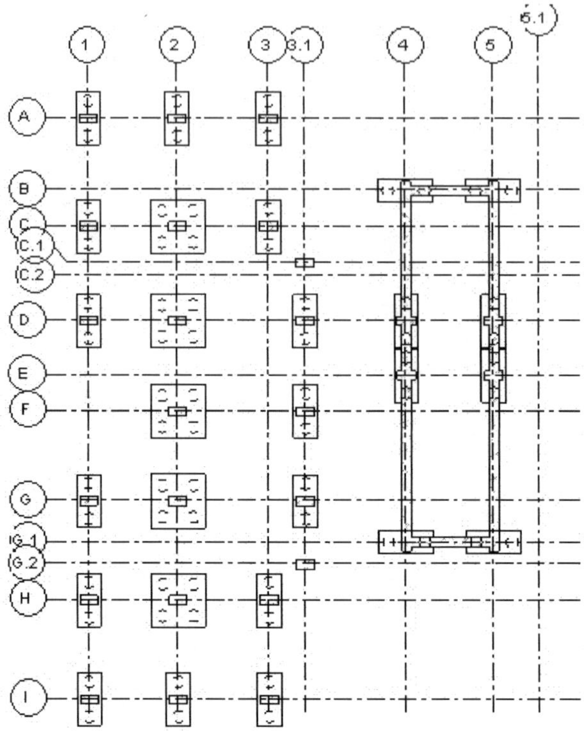

Figure 5-7 *The 2-pile foundations added to the columns*

13. Release the CTRL key, and choose the **Finish** button from the **Multiple** panel of the **Modify | Place Isolated Foundations > At Structural Column** contextual tab. Ignore, if any warning message is displayed. Next, press ESC.

14. Next, choose the **Isolated** tool from the **Foundation** panel of the **Structure** tab; the **Modify | Place Isolated Foundation** contextual tab is displayed.

15. In the **Properties** palette, ensure that the option is selected in the **Type Selector** drop-down list.

    For Imperial **Pile Cap -2 Pile : 31"x61"x35"**
    For Metric **M_Pile Cap -2 Pile : 800 x1500 x900mm**

16. Now place the cursor over the column at the intersection of grid C.1 and grid 3.1; both the grids at the intersection get highlighted.

17. Next, click on the highlighted grids; the selected foundation type is added to the column placed at the intersection of grid C.1 and grid 3.1. Ignore if any warning message is displayed.

18. Repeat steps 16 and 17 to add the foundation type to the column at the intersection of grid G.2 and grid 3.1. Ignore, if any warning message is displayed.

19. After adding the foundations, choose the **Modify** button from the **Select** panel of the **Modify|Place Isolated Foundation** contextual tab to exit the **Isolated** tool.

Adding the Foundation Wall

In this section, you will add the foundation wall to the structural walls in the project.

1. Choose the **Wall** tool from the **Foundation** panel of the **Structure** tab; the **Modify | Place Wall Foundation** contextual tab is displayed.

2. In the **Properties** palette, ensure that the **Wall Foundation : Bearing Footing - 36" x 12"** option in Imperial unit system or **Wall Foundation : Bearing Footing -900 x300mm** in Metric unit system is selected from the **Type Selector** drop-down list.

3. Click on the structural wall located at grid B; the foundation wall of the selected type is added throughout the length of the wall, as shown in Figure 5-8.

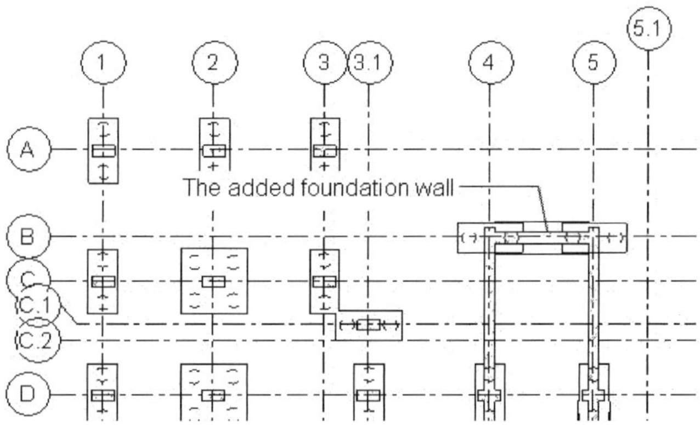

Figure 5-8 *The foundation wall added to the structural wall located at grid B*

4. Now, click on the remaining structural walls in the model; the foundation walls are added to all the structural walls in the project, as shown in Figure 5-9.

5. Choose the **Modify** button from the **Select** panel to exit the **Wall** tool in the **Foundation** panel.

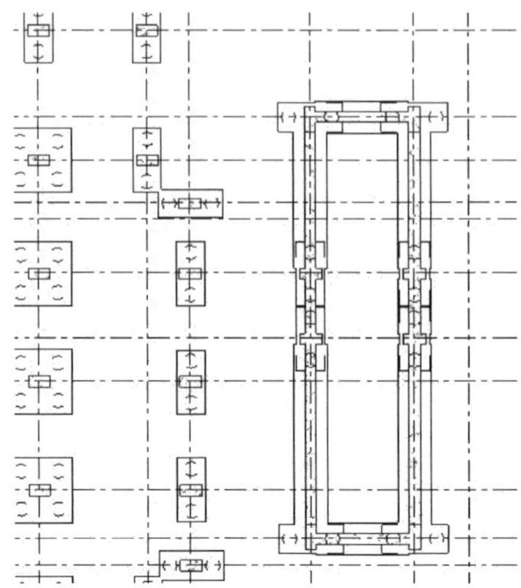

Figure 5-9 The foundation walls added to all the structural walls in the project

Adding the Foundation Slab

In this section, you will add foundation slab to the project.

1. Invoke the **Structural Foundation: Slab** tool from **Structure > Foundation > Slab** drop-down; the **Modify | Create Floor Boundary** contextual tab is displayed.

2. In the **Draw** panel of the **Modify | Create Floor Boundary** contextual tab, ensure that the **Boundary Line** tool is chosen, and then choose the **Rectangle** tool from the list box displayed next to it.

3. In the **Options Bar**, click in the **Offset** edit box and replace the existing value with **6'** (for Imperial) or **1800mm** (for Metric). Next, press ENTER.

4. Move the cursor and place it at the intersection of grid I and grid A and click when the **Midpoint** snap is displayed.

5. Move the cursor down toward the right and place it at the intersection of grid I and grid 3.1, and click when the **Intersection** snap is displayed.

6. Invoke the **Line** tool from the list box in the **Draw** panel of the **Modify | Create Floor Boundary** contextual tab.

7. In the **Options Bar**, enter **3'** in Imperial unit or **900 mm** in Metric unit in the **Offset** edit box to replace the existing value.

8. Move the cursor and place it at the intersection of grid B and the right edge of the rectangle created previously, refer to Figure 5-10.

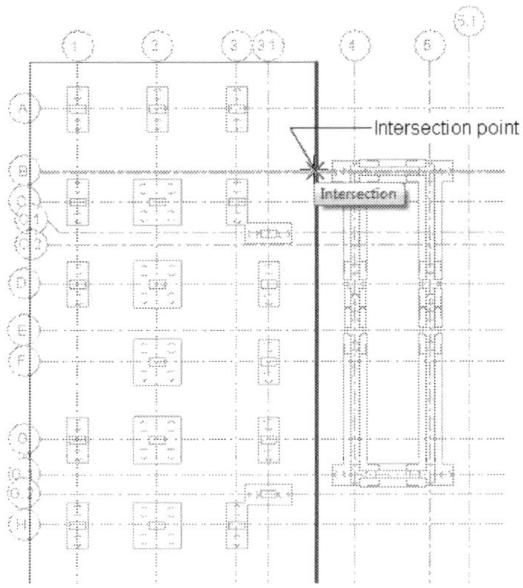

Figure 5-10 *Intersection of grid B and the right edge of the rectangle created previously*

9. Click when grid B and the right edge of the rectangle are highlighted and the **Intersection** snap is displayed.

10. Move the cursor toward the right and place it at the intersection of grid B and grid 5. Click when the **Midpoint and Horizontal** snap is displayed.

11. Move the cursor vertically down and place it at the intersection of grid G.1 and grid 5. Click when the **Midpoint and Vertical** snap is displayed.

12. Move the cursor horizontally left and place it at the intersection of grid G.1 and the right edge of the rectangle created.

13. Click when the **Intersection and Horizontal** snap is displayed. Next, press ESC.

14. In the **Modify** panel of the **Modify | Create Floor Boundary** contextual tab, choose the **Split Element** tool; the cursor changes into a pen symbol.

15. Place the cursor on the point marked **A**, refer to Figure 5-11, and click when the split symbol appears. On doing so, a blue dot is displayed at the point you clicked. Similarly, click at the point marked **B**, refer to Figure 5-11.

16. Press ESC twice and select the split line between the points marked **A** and **B**, refer to Figure 5-11.

17. Press DELETE to remove the split line from the slab boundary.

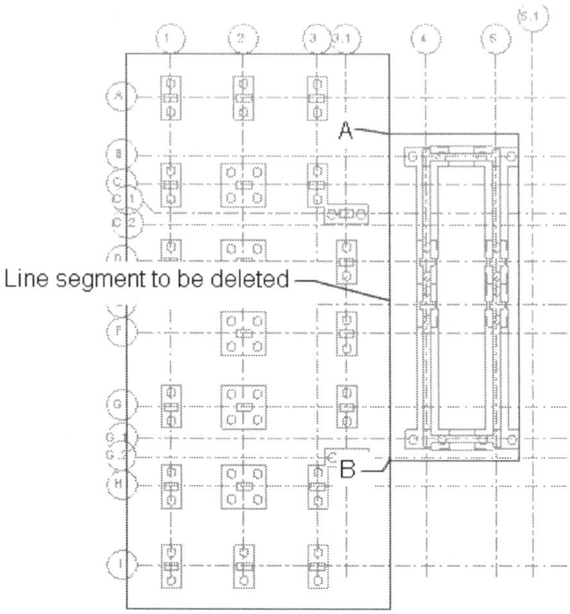

Figure 5-11 Reference points for splitting the slab boundary and the line segment to be deleted

18. Next, choose the **Finish Edit Mode** button from the **Mode** panel of the **Modify | Create Floor Boundary** contextual tab to finish the creation of the boundary of the foundation slab. Note that a warning message may be displayed at the bottom of the screen. Choose **OK** in the warning message box to ignore it.

19. In the **Properties** palette, select the **Foundation Slab : 12" Foundation Slab** option in Imperial unit system or **Foundation Slab : 150 mm Foundation Slab** option for Metric unit system from the **Type Selector** drop-down list. Ignore any warning message, if displayed.

20. Choose the **Modify** button from the **Select** panel of the **Modify | Structural Foundations** contextual tab to create the foundation slab.

Adding Beams

In this section, you will add beams to the **Second Floor** structural plan view of the project.

1. To change the current view to the **Second Floor** structural plan view, double-click on the **Second Floor** sub-node under the **Structural Plans** head in the **Project Browser**.

2. Now, choose the **Beam** tool from the **Structure** panel of the **Structure** tab.

3. In the **Properties** palette, select the **Concrete-Rectangular Beam : 12x24** option for Imperial or **M_Concrete-Rectangular Beam: 300 x 600mm** option for metric from the **Type Selector** drop-down list.

4. In the **Options Bar,** ensure that the **Chain** check box is cleared. Now, move the cursor and place it at the intersection of grid 1 and grid A. Click at the intersection when the **Midpoint** snap is displayed.

5. Move the cursor to the right and place at the center of column. Click at the intersection when the **Midpoint and Horizontal** snap is displayed for the column placed at the intersection of grid A and grid 2. Notice that a beam is added between the intersections of grids.

6. Next, click when the **Midpoint** snap is displayed for the column placed at the intersection of grid 2 and grid A.

7. Move the cursor horizontally right and place it at the center point of the column placed at intersection point of grid A and grid 3. Click at the intersection when the **Midpoint and Horizontal** snap is displayed.

8. Repeat the procedure followed in steps 6 and 7 to add the remaining beams to the structural plan view. Refer to Figure 5-12 for the placement of beams. Note that while adding beams, a warning message may be displayed at the bottom of the screen. Choose the **OK** button to ignore the warning message box and close it.

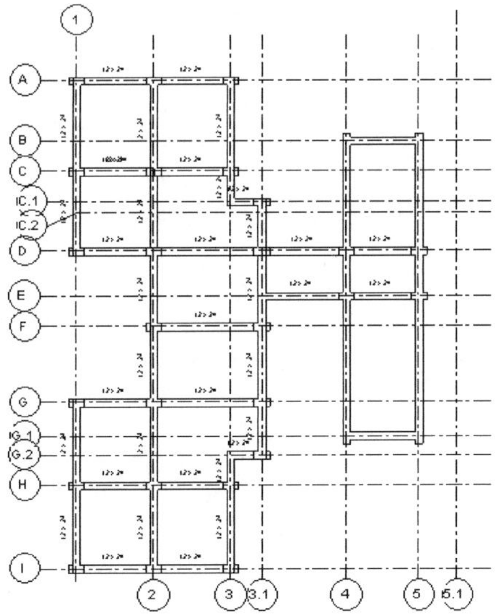

Figure 5-12 *The beams and other structural elements in the* ***Second Floor*** *structural plan view*

Note

While referring to Figure 5-12 for the placement of beams, notice the tags on the beams to distinguish between beams and structural walls.

Tip

In case, you want to show the tags of the beams then choose the ***Tag on Placement*** *button from the* ***Tag*** *panel of the* ***Modify | Place Beam*** *contextual tab.*

9. Choose the **Modify** button from the **Select** panel of the **Modify |Place Beam** contextual tab to exit the **Beam** tool.

Modifying Beams

In this section, you will select all the instances of the beams in the project view and modify their instance properties.

1. Click on the beam placed at grid A between grid 1 and grid 2, and then right-click; a shortcut menu is displayed.

2. Choose **Select All Instances > Visible in View** from the shortcut menu, as shown in Figure 5-13; all beams in the project view area are selected.

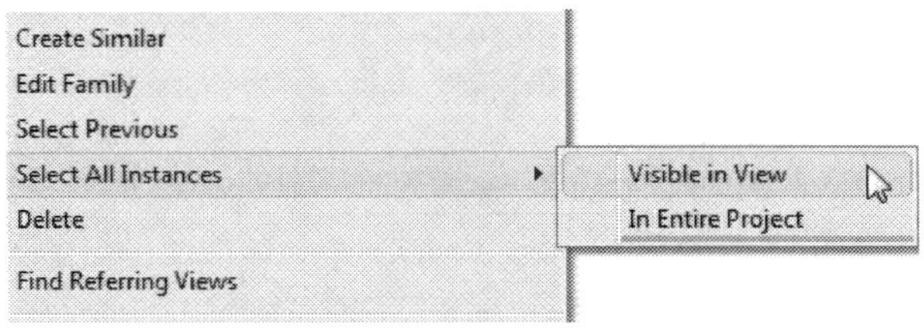

Figure 5-13 *Choosing* ***Select All Instances > Visible in View*** *from the shortcut menu*

3. Now, press SHIFT key and then select the beams created on grids 4,5, B and G.1; these beams are removed from the selection.

4. In the **Properties** palette, specify the instance parameters as:

For Imperial	Start Level Offset = **-1'**	End Level Offset = **-1'**
For Metric	Start Level Offset = **-300mm**	End Level Offset = **-300mm**

5. Next, choose **Apply** in the **Properties** palette and then press ESC.

Adding a Structural Floor

In this section, you will add a structural floor to the project in the **Second Floor** structural plan view.

1. Ensure that the **Second Floor** sub-node under the **Structural Plans** node is highlighted in the **Project Browser**.

2. Invoke the **Floor: Structural** tool from the **Structure > Structure > Floor** drop-down.

3. In the **Properties** palette, select the **Generic 12"** option from the **Type Selector** drop-down list or **Generic 300mm** option in Metric unit system.

4. In the **Modify | Create Floor Boundary** contextual tab, ensure that the **Boundary Line** option is chosen in the **Draw** panel.

5. Choose the **Line** tool from the list box displayed right next to the **Boundary Line** tool.

6. In the **Options Bar**, ensure that the **Chain** check box is selected and then in the **Offset** edit box, replace the default value.
   For Imperial **1'**
   For Metric **300 mm**

7. Move the cursor and place it at the intersection of grid A and grid 1 and click when the **Endpoint** snap is displayed.

8. Move the cursor toward the right and place it at the intersection of grid A and grid 3. Click when the **Endpoint and Horizontal** snap is displayed.

9. Click on the other grid intersections in the following order:
   (C.1, 3); (C.1, 3.1); (D, 3.1); (D, 4); (C.1, 4); (C.1, 5); (F, 5); (F, 4); (E, 4); (E, 3.1); (G.2, 3.1); (G.2, 3); (I, 3); (I, 1); (G,1); (G, 2); (D, 2); (D, 1); and (A, 1)

Note
The grid intersection points are denoted by their intersecting grid lines. For example, (C.1, 3.1) represents the intersection point of grid C.1 and grid 3.1

After you click on the last grid intersection, the floor boundary is created, as shown in Figure 5-14.

10. Choose the **Modify** tool from the **Select** panel to exit the current tool.

11. Choose the **Trim/Extend to Corner** tool from the **Modify** panel and then select the recently created line and the horizontal line starting from grid 1; the two lines get joined.

12. Next, choose the **Finish Edit Mode** button from the **Mode** panel of the **Modify | Create Floor Boundary** tab; the **Revit** message box is displayed.

13. Choose the **Modify** button from the **Select** panel of the **Modify | Floors** contextual tab to complete the floor creation.

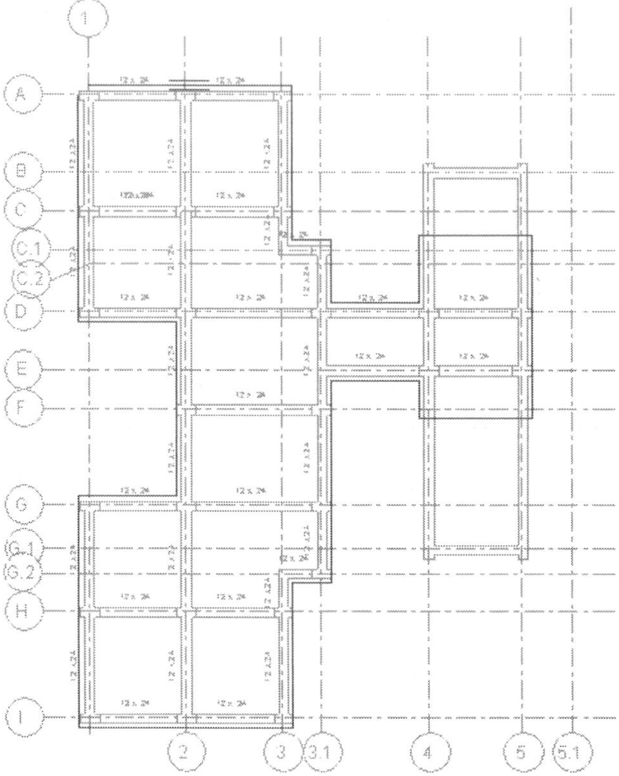

Figure 5-14 *Structural plan displaying the floor boundary*

Saving the Project

In this section, you will save the project file using the **Save As** tool.

1. To save the project, choose the **Save As > Project** option from the **File** menu; the **Save As** dialog box is displayed.

2. In this dialog box, browse to the *C:\rst_2020* location and create a folder with name *c05_rst_2020_tut*. Open the *c05_rst_2020 _tut* folder and enter **c05_Commercial-Complex_tut1** for Imperial and **M_c05_Commercial-Complex_tut1** for Metric in the **File name** edit box.

3. Choose the **Save** button; the **Save As** dialog box closes and the project file is saved.

Closing the Project

1. To close the project, choose the **Close** option from the **File** menu.

The file is closed and this completes Tutorial 1 of Chapter 5.

EXERCISE

Exercise 1 Academic Institution

In this exercise, you will add foundations, beams, and structural floor to the project file created for Exercise 1 of Chapter 4, refer to Figures 5-15 and 5-16. Use the following project parameters and specifications: **(Expected time: 1 hr 15 min)**

1. Foundations: (To be added in the **Entry** structural plan view)
   For Imperial Type: **Wall Foundation Bearing Footing : 36"x12"**
   Footing Rectangular : 72"x48"x18"
   For Metric Type: **Wall Foundation Bearing Footing : 900 x 300**
   M_Footing - Rectangular : 1800 x1200 x450mm

2. Beams: (To be added in the **Second Floor** structural plan view)
   For Imperial Type: **Concrete-Rectangular Beam : 12x24**
   For Metric Type: **M_Concrete-Rectangular Beam : 300 x600**

3. Structural Floor:
   For Imperial Type: **3" LW Concrete on 2" Metal Deck**
   1 1/2" Metal Roof Deck
   For Metric Type: **160mm LW Concrete on 50mm Metal Deck**
   75mm Metal Roof Deck

4. Project file to be used:
   For Imperial *c04_Academic-Institution_exer1.rvt.*
   For Metric *M_c04_Academic-Institution_exer1.rvt.*

5. File name to be assigned:
   For Imperial *c05_Academic-Institution_exer1.rvt.*
   For Metric *M_c05_Academic-Institution_exer1.rvt.*

6. For location and placement of beams, foundations and other structural elements, refer to Figure 5-15 and Figure 5-16.

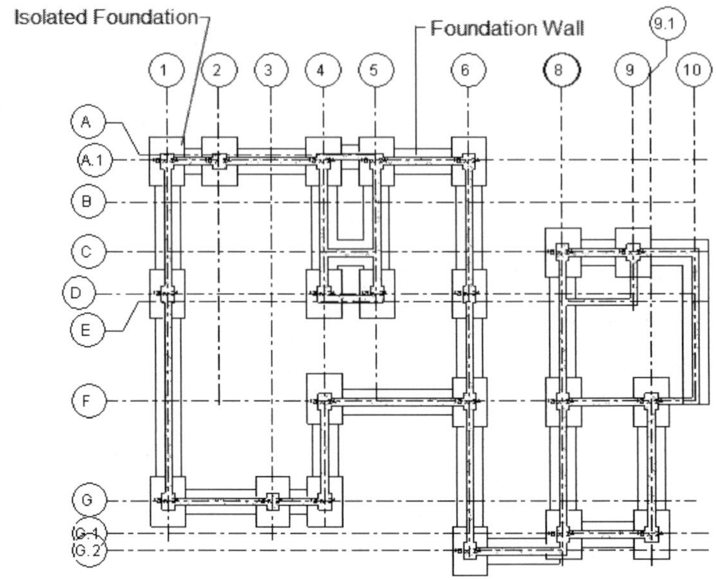

Figure 5-15 *Structural plan view at the* ***Entry*** *level*

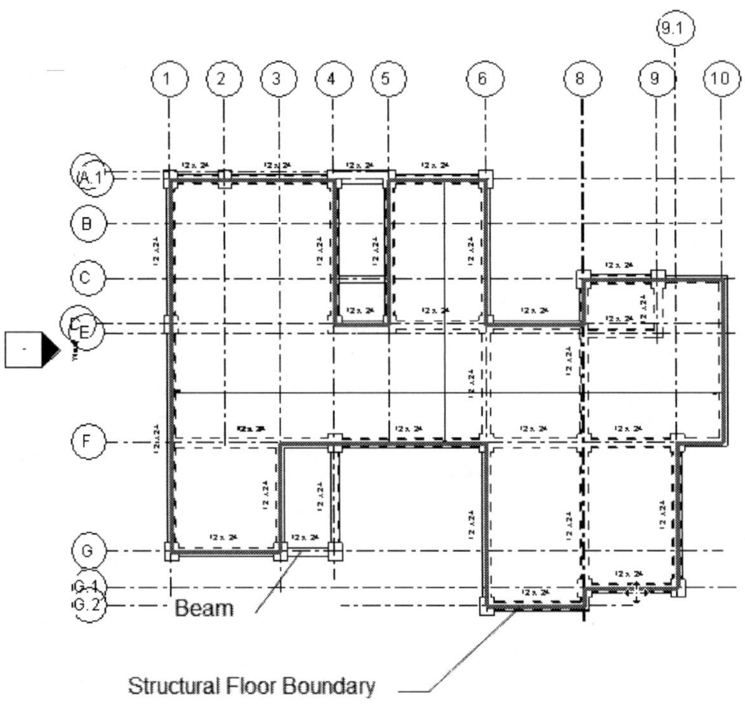

Figure 5-16 *Structural plan view at the* ***Second Floor*** *level*

Chapter 6

Editing Tools

Learning Objectives

After completing this chapter, you will be able to:

- *Create a selection set of elements*
- *Move and copy elements*
- *Use the Cut and Paste tools*
- *Understand the use of the Rotate and Mirror tools*
- *Create an array of elements using the Array tool*
- *Use the Match, Align, Trim, Offset, and Delete tools*

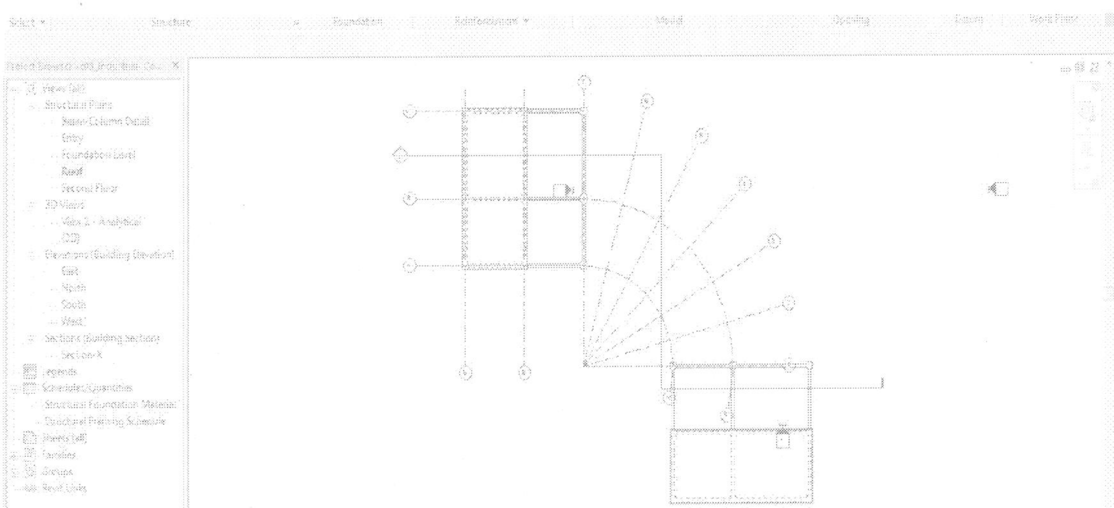

In the previous chapters, you have learned to create structural elements such as structural walls, columns, and framing elements. In this chapter, you will learn about the editing process and the tools that are used to modify elements based on the design requirement. Some of the editing processes such as modifying type and instance properties, flipping orientation, and many more have already been discussed in the previous chapters. Also, you will learn to select and edit elements by using the object controls and editing tools such as **Move**, **Copy**, **Trim/Extend**, **Delete**, **Rotate**, **Mirror**, **Array**, **Align**, **Match**, and so on.

SELECTING ELEMENTS

Selecting elements in a drawing is an important and regular task needed for editing elements. Therefore, it is important to learn the basic and advanced methods of selection. Learning these methods will help you speed up the selection process as well as reduce the errors that may occur due to the wrong selection of elements.

In Revit Structure, you need to select the element or group of elements for editing. There are several methods that can be used to select the elements from a drawing. The method used to select an element may differ depending on the nature of selection. For instance, you may need to select a single element, multiple elements of different categories, multiple elements of same categories, and so on.

Selecting Multiple Elements

The methods of selecting multiple elements are mentioned next.

Using the CTRL Key
Using the Selection Window
Using the TAB Key (For Walls)
Using the Select All Instances Option
Isolating Elements Using the Selection Box

MODIFYING TOOLS

In Revit, there are various tools that can be used to edit the sketched entities. These tools are used to trim, move, copy extend, offset, or mirror the sketched entities. You can also perform various other editing operations by using these tools. Various editing operations and the tools used to perform them are:

Move and copy elements
Trim and Extend elements
Cut and paste elements
Rotate elements
Mirroring elements
Offset Elements
Array elements
Matching Elements
Aligning elements and working with constraints
Deleting elements
Splitting elements

Group Elements
Pinning and unpinning elements
Scaling elements
Measuring Distance Between References and Along an Element

TUTORIAL

Tutorial 1 Commercial Complex

In this tutorial, you will rotate some of the structural foundations and columns of the *Commercial Complex* project created in Tutorial 1 of Chapter 5. Further, you will copy beams, columns, and floor in the **Second Floor** structural plan view and paste them on the **Roof** level, using the **Filter** tool. Use the following project parameters and specifications:

(Expected time: 45 min)

1. Rotation Angle: **90°**
2. Elements to be copied to level: columns, beams, and structural floor
3. File name to be assigned:

   For Imperial *c06_Commercial-Complex_tut1.rvt*
   For Metric *M_c06_Commercial-Complex_tut1.rvt*

The following steps are required to complete this tutorial:

a. Open Tutorial 1 of Chapter 5.
b. Rotate structural foundations (at the **Entry** level).
c. Rotate columns (at the **Entry** level).
d. Copy and paste columns, floors, and beams.
e. Save the project using the **Save As** tool.
f. Close the project using the **Close** tool.

Opening the Existing Project

In this section, you will open the project file created in Tutorial 1 of Chapter 5.

1. Choose **Open > Project** from the **File** menu; the **Open** dialog box is displayed.

2. In this dialog box, browse to *C:\rst_2020\c05_rst_2020_tut* location and then choose the *c05_Commercial-Complex_tut1.rvt* for Imperial *or M_c05_Commercial-Complex_tut1.rvt* for Metric project file. You can also download this file from *https://www.cadsofttech.com*. The path of the file is as follows: *Textbooks > Civil/GIS > Revit Structure > Revit Structure 2020 for Novices.*

3. Choose the **Open** button from the **Open** dialog box; the selected project file opens in the drawing window in the **Second Floor** structural plan view.

Rotating the Structural Foundations

In this section, you will select some of the structural foundations in the **Entry** project view and rotate it.

1. Double-click on the **Entry** node in the **Structural Plans** head in the **Project Browser** to open the corresponding structural floor plan view.

2. Place the cursor on the structural foundation located at the intersection of grid 3.1 and grid C.1; the structural foundation is highlighted, as shown in Figure 6-1.

3. Click on the highlighted foundation to select it.

4. Choose the **Rotate** tool from the **Modify** panel of the **Modify | Structural Foundations** contextual tab; a dashed preview box is displayed around the selected foundation.

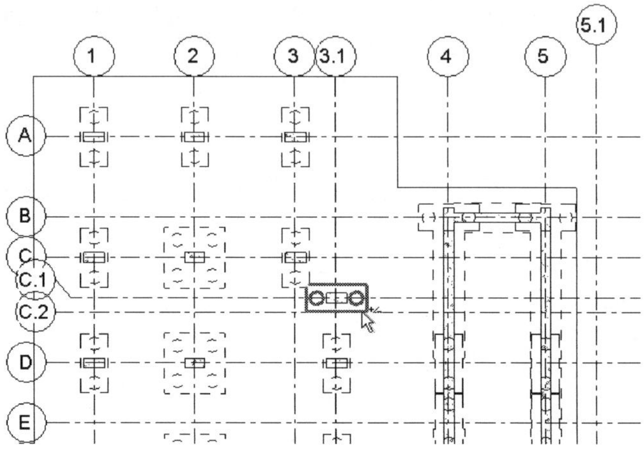

Figure 6-1 *The structural foundation highlighted*

5. In the **Options Bar**, click on the **Angle** edit, enter **90** and then press ENTER; the selected structural foundation is rotated by 90 degrees about its center, refer to Figure 6-2.

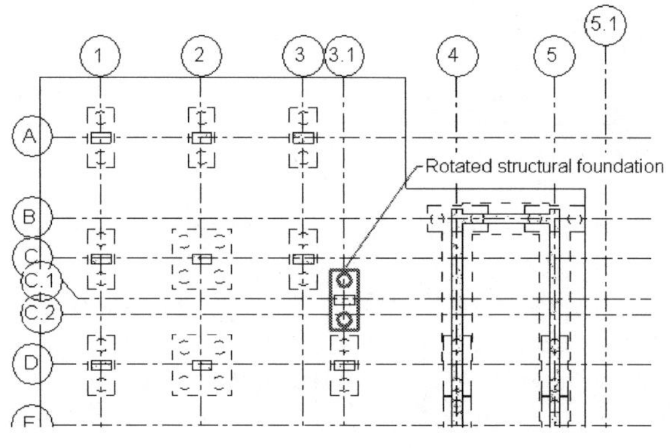

Figure 6-2 *The structural foundation rotated*

6. Place the cursor on the structural foundation located at the intersection of grids 3.1 and G.2. Click to select the highlighted structural foundation.

7. Repeat the procedure followed in steps 4 and 5 to rotate the selected structural foundation.

8. Choose the **Modify** tool from the **Select** panel to exit the selection.

Rotating Columns

In this section, you will rotate some of the structural columns in the structural model.

1. Place the cursor over the column located at the intersection of grid 4 and grid D; the column is highlighted, as shown in Figure 6-3. Click on the highlighted column to select it.

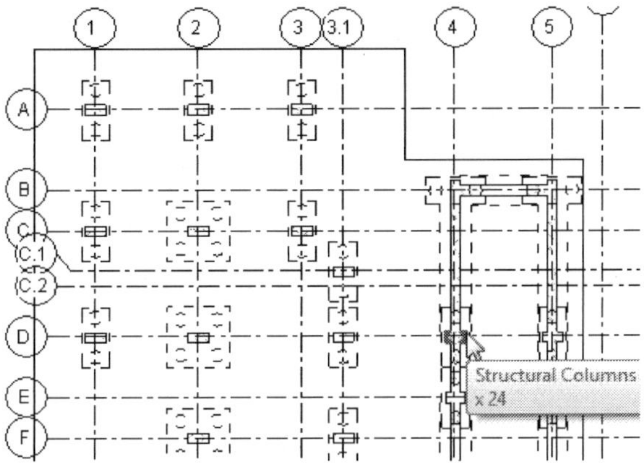

Figure 6-3 *The structural column highlighted*

2. Choose the **Rotate** tool from the **Modify** panel of the **Modify | Structural Columns** contextual tab; a dashed preview box is displayed around the structural column.

3. In the **Options Bar**, click in the **Angle** edit box and enter **90**. Press ENTER; the selected column is rotated 90 degrees about its center.

4. Next, repeat the procedure followed in steps 1 to 3 to rotate the columns located at the following grid intersections: (4, E); (5, D); and (5, E). After rotating the desired columns, the structural plan view at the **Entry** level appears as shown in Figure 6-4.

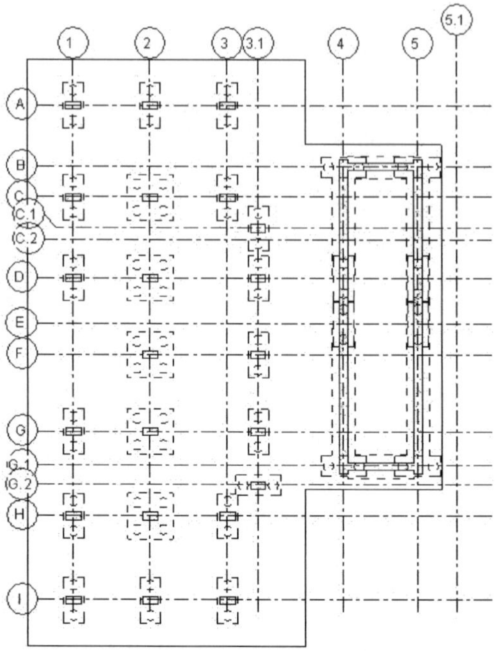

*Figure 6-4 The structural plan view at the **Entry** level*

Copying and Pasting Structural Columns, Floors, and Beams

In this section, you will select structural columns and beams in the **Second Floor** structural plan view and then copy and paste them at the **Roof** level. To select the beams and columns, you will use the **Filter** tool.

1. Double-click on **Second Floor** in the **Structural Plans** head in the **Project Browser** to open the corresponding structural floor plan view.

2. Place the cursor between grids A and 1, as shown in Figure 6-5.

3. Now, press and hold the left mouse button and drag the cursor diagonally down and place it between grid I and grid 5.1, as shown in Figure 6-6. On dragging the cursor down, a selection window emerges and all the elements including grids enclosed inside the selection box are highlighted.

4. Release the left mouse button; all the elements in the selection box are selected.

5. Choose the **Filter** tool from the **Selection** panel of the **Modify | Multi-Select** contextual tab; the **Filter** dialog box is displayed.

6. In the **Filter** dialog box, choose the **Check None** button; all the check boxes in the **Category** column are cleared.

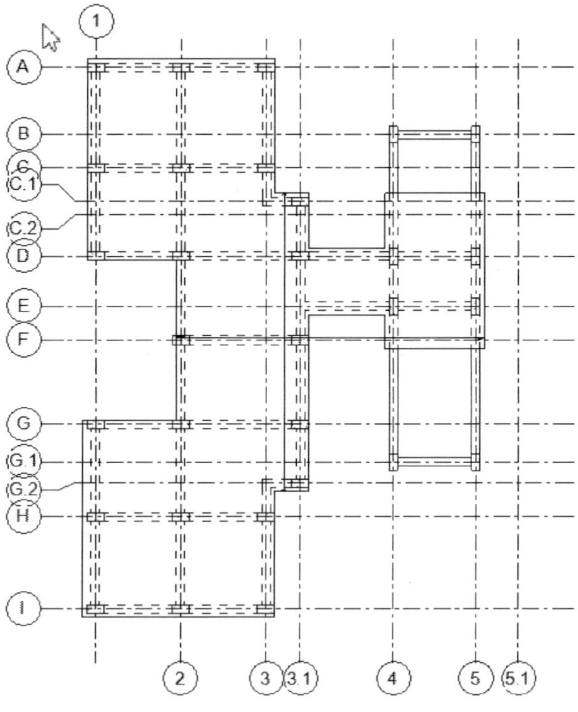

Figure 6-5 *Placing the cursor to select the start point of the selection window*

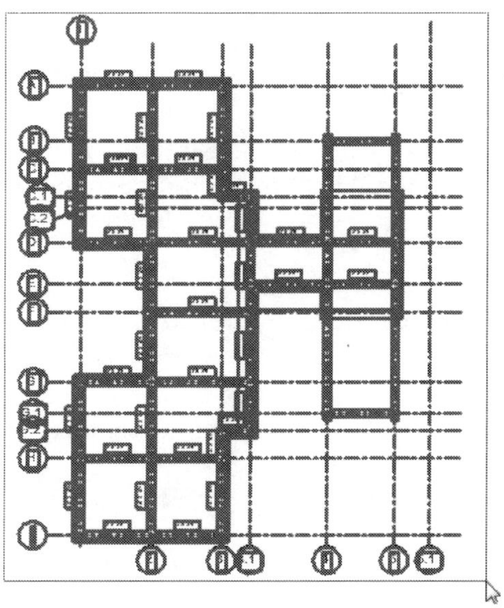

Figure 6-6 *Placing the cursor to select the endpoint of the selection window*

7. Now, select the following check boxes in the **Category** column: **Floors**, **Structural Columns**, **Structural Framing (Girder)**, and **Structural Framing (Other)**.

8. Choose **Apply** and then **OK**; the **Filter** dialog box closes.

9. Invoke the **Copy to Clipboard** tool from the **Clipboard** panel of the **Modify | Multi-Select** contextual tab; the selected elements are copied to the clipboard.

10. Next, choose the **Aligned to Selected Levels** tool from **Modify | Multi-Select > Clipboard > Paste** drop-down; the **Select Levels** dialog box is displayed.

11. Click on the **Roof** option in this dialog box and then choose **OK**; the selected elements are copied to the **Roof** level. Ignore warning message, if any.

12. Choose the **Modify** tool from the **Select** panel to exit the selection.

Adding Beams to the Roof Level

In this section, you will add beams to the columns that are projected to the **Roof** level.

1. Double-click on the **Roof** node under the **Structural Plans** head in the **Project Browser**.

2. Invoke the **Beam** tool from the **Structure** panel in the **Structure** tab.

3. In the **Properties** palette, ensure that the **Concrete-Rectangular Beam: 12 x 24** (**M_Concrete-Rectangular Beam: 300x600** for metric) option is selected in the **Type Selector** drop-down list.

4. Now, using the **Beam** tool that is already invoked, add beams between the following grid intersections:
   (G.1,4) and (E,4); (E,4) and (D,4); (D,4) and (B,4); (B,4) and (B, 5); (B,5) and (D, 5); (D,5) and (E, 5); (E,5) and (G.1, 5); and (G.1,5) and (G.1, 4)

5. Next, choose the **Modify** tool from the **Select** panel and then select the beams created in the previous step by using the CTRL key, and in the **Properties** palette, replace the existing values in the value fields of the **Start Level Offset** and **End Level Offset** parameters with **-1'** for Imperial **(-300mm** for Metric).

6. Choose the **Modify** tool to exit from the current selection.

Saving the Project

In this section, you will save the project file using the **Save As** tool.

1. Choose the **Save As > Project** option from the **File** menu; the **Save As** dialog box is displayed.

2. In this dialog box, browse to the *C:\rst_2020* location and create a folder with name **c06_rst_2020_tut**. Next, open the created folder and enter **c06_Commercial-Complex_tut1** for Imperial and **M_c06_Commercial-Complex_tut1** for Metric unit system in the **File name** edit box.

3. Choose the **Save** button; the **Save As** dialog box closes and the project file is saved.

Closing the Project

1. Choose the **Close** option from the **File** menu.

The file is closed and this completes Tutorial 1 of Chapter 6.

EXERCISE

Exercise 1 Academic Institution

In this exercise, you will mirror the structural model created in Exercise 1 of Chapter 5. You will define grid 10 as a mirror axis. Further, you will copy the structural floor, beams, and columns to the **Roof** level. Use the following project parameters and specifications:

(Expected time: 1 hr)

1. Use the **Second Floor** structural plan view to mirror the floor and beams.
2. Use the **Entry** structural plan view to mirror the foundations.
3. File name to be assigned: *c06_Academic-Institution_exer1.rvt.*
4. Copy the structural floor, beams, and columns displayed in the **Second Floor** view and paste it at the **Roof** level.
5. For reference of the mirror line, refer to Figures 6-7 and 6-8.

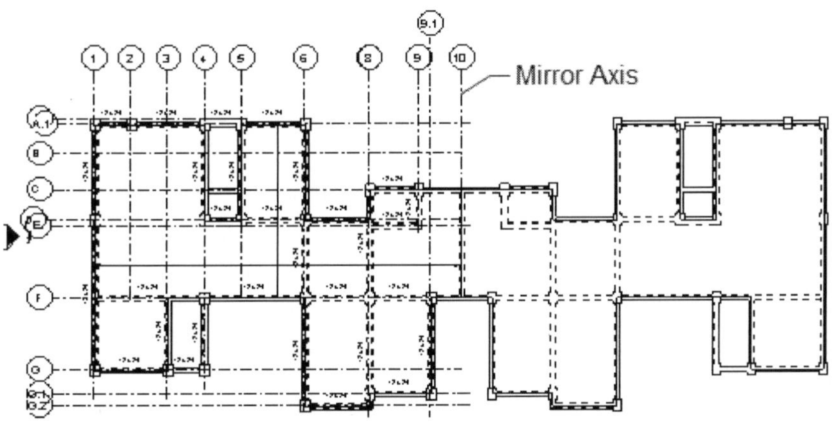

Figure 6-7 *The mirrored structural elements at the* ***Second Floor*** *level*

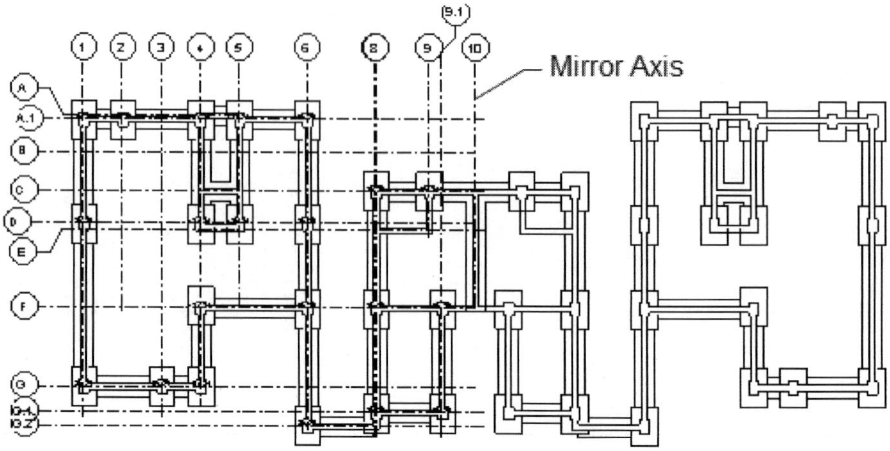

Figure 6-8 *The mirrored structural elements at the* ***Entry*** *level*

Chapter 7

Documenting Models and Creating Families

Learning Objectives

After completing this chapter, you will be able to:

- *Add permanent dimensions*
- *Convert temporary dimensions into permanent dimensions*
- *Modify a dimension and its parameters*
- *Add tags to project*
- *Understand family creations*

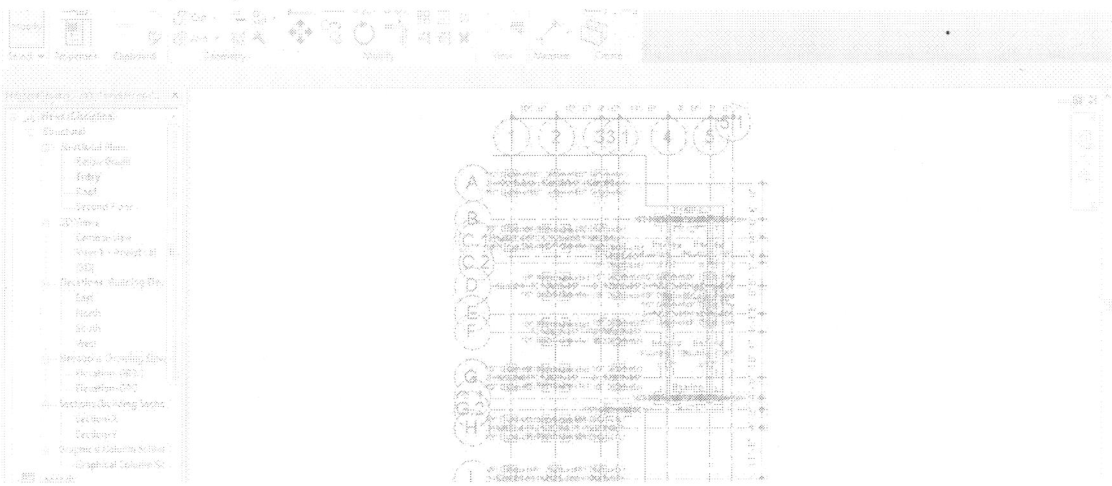

INTRODUCTION

In the previous chapters, you learned how to add structural grids, columns, framing members, and various other structural elements to a project. In this chapter, you will learn how to document a project and prepare it as a working drawing by using various tools and techniques in Revit. To document a model, you need to place dimensions to the structural plan view, add text notes and place tags to columns, beams, and structural elements wherever required. In this chapter, you will also learn about various dimensioning terms and tools, tools for placing text notes, and placing tags and beam annotations in your project.

DIMENSIONING

Dimensions play a crucial role in the presentation of a project. Although the structural model of a project can convey what the appearance or design of the model would be, yet to materialize it at the site, there must be information and statistics regarding each element involved in the model. Since the design of a project is used for the actual construction of the project, it is essential to describe each element of the project in terms of actual measurement parameters such as length, width, height, angle, radius, diameter, and so on. All these information can be added to a project by using dimensions.

Types of Dimension

In Revit, you can use two types of dimension: temporary and permanent. By default, in the temporary and permanent dimensions, the units that are specified in the initial start-up of a project are used as the default units. Unlike temporary dimensions, the permanent dimensions are view-specific. It means if you change the view of a project, the permanent dimensions will not be visible.

Using Temporary Dimensions

Revit displays temporary dimensions around the component or element when you place or sketch it in a view. The dimension that appears dynamically while drawing or placing an element is called temporary dimension. This type of dimension is not view-specific and appears in any view while drawing or selecting an element. Temporary dimensions help you position elements at the desired location and references. While sketching the lines of desired length at the desired angle instantly, temporary dimensions can help you speed up your drafting work.

Adding Permanent Dimensions

Permanent dimensions are added for a specific measurement. In Revit, you can access various dimension tools from the **Dimension** panel in the **Annotate** tab, as shown in Figure 7-1. You can choose appropriate dimension type and dimension tool to add dimensions to an element.

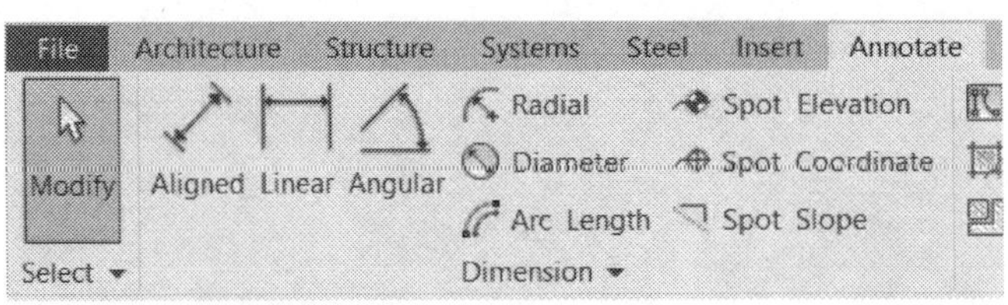

***Figure 7-1** Various dimensioning tools in the **Dimension** panel of the **Annotate** tab*

TEXT NOTES

Text notes are important parts of a project detail. They not only help in adding the specification of various structural elements but also in conveying the specific design intent. Revit provides a variety of options to add text notes to a structural model detail view by using the **Text** tool.

Adding Text Notes

Ribbon:	Annotate > Text > Text

In Revit, text notes are view-specific entities. You can add different text notes to different views. Generally, text notes are added to the plan view, section view, callout view, or in drafting view. The selection of a view in which the text notes are added depends on the project requirement. To add a text note to a desired view, invoke the **Text** tool from the **Text** panel; the **Modify | Place Text** contextual tab will be displayed, as shown in Figure 7-2.

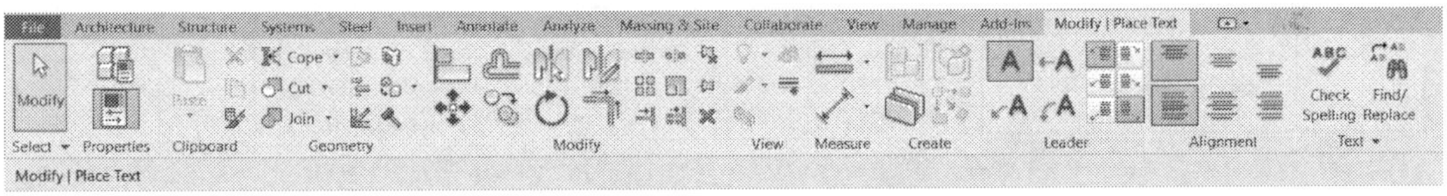

Figure 7-2 *Various options in the **Modify / Place Text** tab*

CREATING FAMILIES

Revit family is defined as a collection of parametric components or elements having similar graphical representation, utility, and common parameters. By nature, all the elements used in Revit are family-based. A Revit family has a wide range of collections starting from a wall to a window and extends up to two-dimensional annotations or symbols required for your project. In Revit, each family inherits the character of a specific type. A type is a member of Revit family that consists of certain parameters known as **Type Parameters**. These parameters are constant for identical types. For example, the **C-Channel-Column** family can have different types such as **C-Channel-Column C12x30**, **C-Channel-Column C15x40**, and others. You can access these types from the **Type Selector** drop-down list.

TUTORIAL

Tutorial 1 — Commercial Complex

In this tutorial, you will open the *Commercial Complex* project created in Tutorial 1 of Chapter 6 and add dimensions to it in different structural plan views. Further, you will annotate beams and columns using various annotation tools. Use the following project parameters:

(Expected time: 25 min)

1. File name to be assigned:
   For Imperial *c07_Commercial-Complex_tut1.rvt*
   For Metric *M_c07_Commercial-Complex_tut1.rvt*

The following steps are required to complete this tutorial:

a. Open the project file.
   For Imperial *c06_Commercial-Complex_tut1.rvt*
   For Metric *M_c06_Commercial-Complex_tut1.rvt*
b. Dimension grid lines.
c. Add the foundation, column, and slab tags.
d. Add beam annotations.
e. Save the project using the **Save As** tool.
f. Close the project using the **Close** tool.

Opening the Existing Project

In this section, you will open the project file created in the Tutorial 1 of Chapter 6.

1. Choose **Open > Project** from the **File** menu; the **Open** dialog box is displayed.

2. In this dialog box, browse to *C:/revit_2020_structural_projects/c06* location and choose the *c06_Commercial-Complex_tut1.rvt* for Imperial and *M_c06_Commercial-Complex_tut1.rvt* for Metric project file. You can also download this file from *https://www.cadsofttech.com*. The path of the file is as follows: *Textbooks > Civil/GIS > Revit Structure > Revit Structure 2020 for Novices*.

3. Choose the **Open** button from the **Open** dialog box; the selected project file opens in the drawing window in the **Second Floor** structural plan view.

Dimensioning the Grid Line

In this section, you will add dimensions to the grid lines displayed in the **Entry** structural plan view.

1. Double-click on the **Entry** node in the **Structural Plans** head in the **Project Browser** to open the corresponding structural floor plan view.

2. Choose the **View** tab and then invoke the **Visibility/ Graphics** tool from the **Graphics** panel of this tab; the **Visibility/Graphic Overrides for Structural Plan: Entry** dialog box is displayed.

3. In this dialog box, choose the **Annotation Categories** tab if not selected and then select the **Dimensions** check box displayed if it is not selected under the **Visibility** column.

4. Choose the **OK** button; the **Visibility/Graphic Overrides for Structural Plan: Entry** dialog box is closed.

5. Choose the **Annotate** tab and then choose the **Aligned** tool from the **Dimension** panel.

6. In the **Properties** palette, ensure that the **Linear Dimension Style : Arrow - 3/32" Arial** option for Imperial or the **Linear Dimension Style : Arrow -2.5mm Arial** option for Metric is selected from the **Type Selector** drop-down list. Next, choose the **Edit Type** button; the **Type Properties** dialog box is displayed.

7. In this dialog box, choose the **Duplicate** button; the **Name** dialog box is displayed.

8. In the **Name** edit box of the dialog box, delete the existing text and enter **Small Dot- 3/32" Arial** for Imperial or enter **Small Dot- 2.5mm Arial** for Metric and then choose the **OK** button; the **Name** dialog box is closed.

9. In the **Type Properties** dialog box, click in the value field of the **Tick Mark** parameter; a drop-down list is displayed. Next, select the desired option from this drop-down list.
   For Imperial **Filled Dot Small**
   For Metric **Filled Dot 2mm**

10. Click in the value field of the **Centerline Symbol** parameter; a drop-down list is displayed. Select the **Centerline** option for imperial and for metric select **M_Centerline** from the drop-down list.

11. Choose the **OK** button; the **Type Properties** dialog box is closed. Also, notice that in the drawing area, the dimension symbol appears with the cursor.

12. Move the cursor above grid A and place it over grid 1; the grid 1 gets highlighted.

13. Click and then move the cursor to the right and place it on grid 2. Next, click when grid 2 gets highlighted. Notice that a dimension between grid 1 and grid 2 is displayed, refer to Figure 7-3.

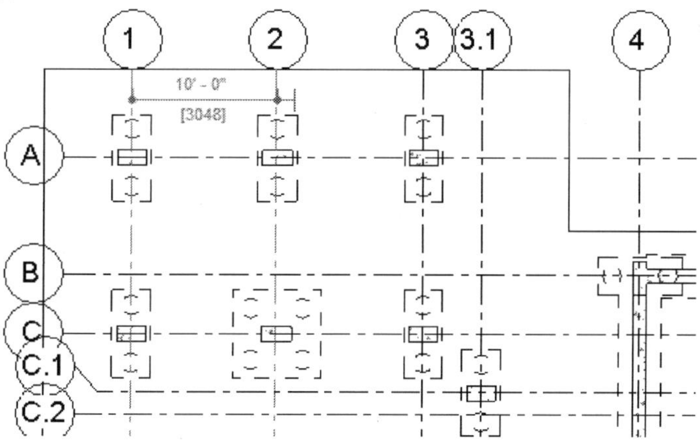

Figure 7-3 *The dimension created between grid 1 and grid 2*

14. Move the cursor over grid 3 and click when it gets highlighted.

15. Using the procedure followed in step 14, provide the dimensions till grid 5.1. After you have clicked over the grid 5.1, move the cursor up and place it above the grid bubbles, as shown in Figure 7-4.

16. Now click at the desired location, refer to Figure 7-4; the dimensions are added. Note that you can place the dimensions based on your requirement.

17. Repeat the procedures followed in steps 12 to 16 and add dimensions for all the horizontal grids from A to I. The dimensions are placed on the right of the structural plan view, refer to Figure 7-5.

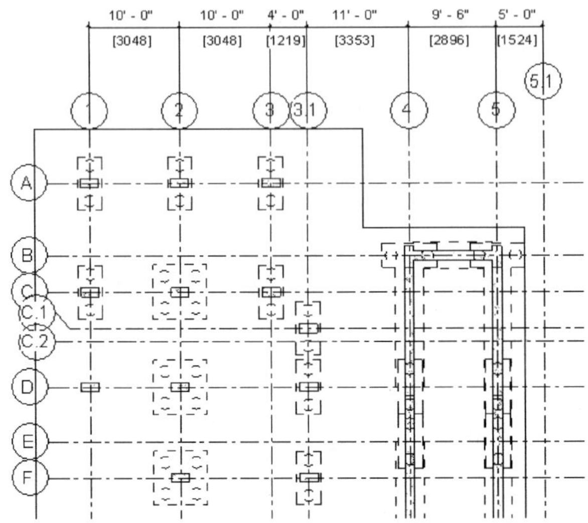

Figure 7-4 *Dimensions placed at a location above grid bubbles*

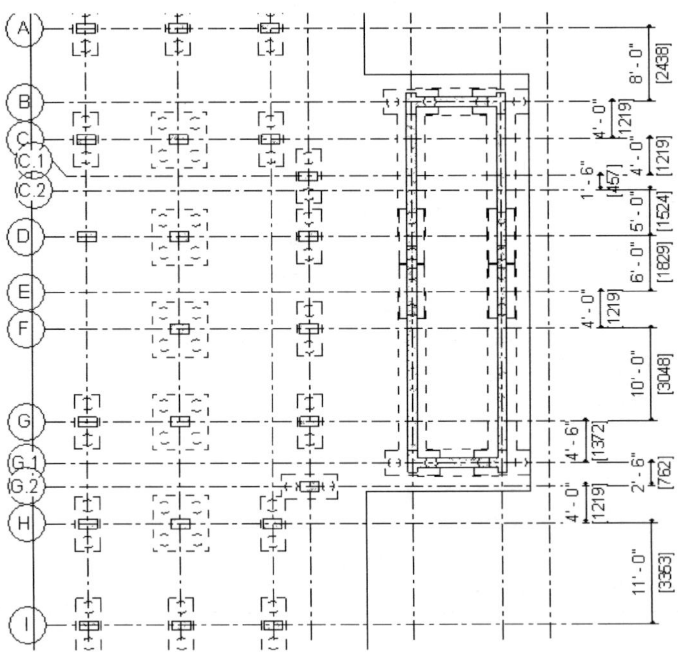

Figure 7-5 *The dimensions placed at a location on the right of grid bubbles*

18. Press ESC twice to exit the tool.

Adding Tags to Foundations, Columns, and the Slab

In this section, you will add tags to foundations, columns, and the slab in the **Entry** structural plan view.

1. Ensure that the **Annotate** tab is chosen and then invoke the **Tag All** tool from the **Tag** panel; the **Tag All Not Tagged** dialog box is displayed.

2. Select the check box corresponding to the **Structural Foundation Tags** category in the **Category** column.

3. Next, select the check box corresponding to the **Structural Column Tags** category with the **Structural Column Tag** as the loaded tag; the selected category is highlighted, as shown in Figure 7-6.

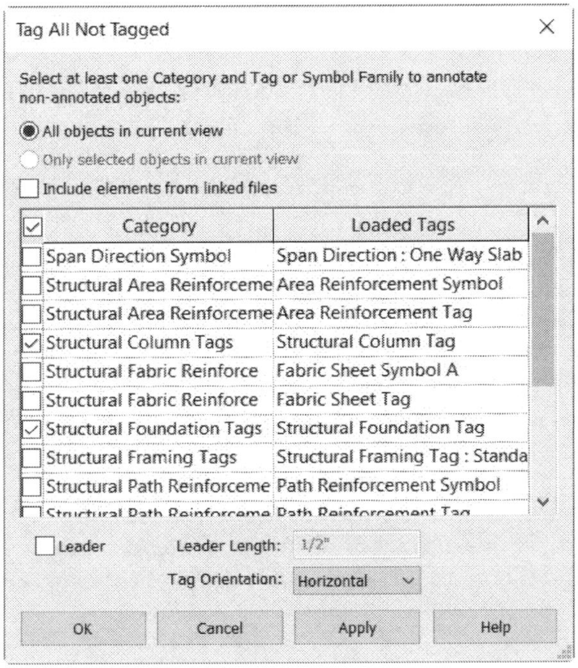

Figure 7-6 *The **Tag All Not Tagged** dialog box with selected categories highlighted*

4. Choose the **Apply** button and then the **OK** button; the **Tag All Not Tagged** dialog box is closed and the selected tag categories are displayed in the **Entry** structural plan view.

5. Next, choose the **Tag by Category** tool from the **Tag** panel and place the cursor over the structural foundation slab; the structural foundation slab is highlighted.

6. Click on the highlighted slab when a preview of the tag appears; the foundation slab tag is added to the foundation slab, as shown in Figure 7-7.

7. Press the ESC key twice to exit the selection mode.

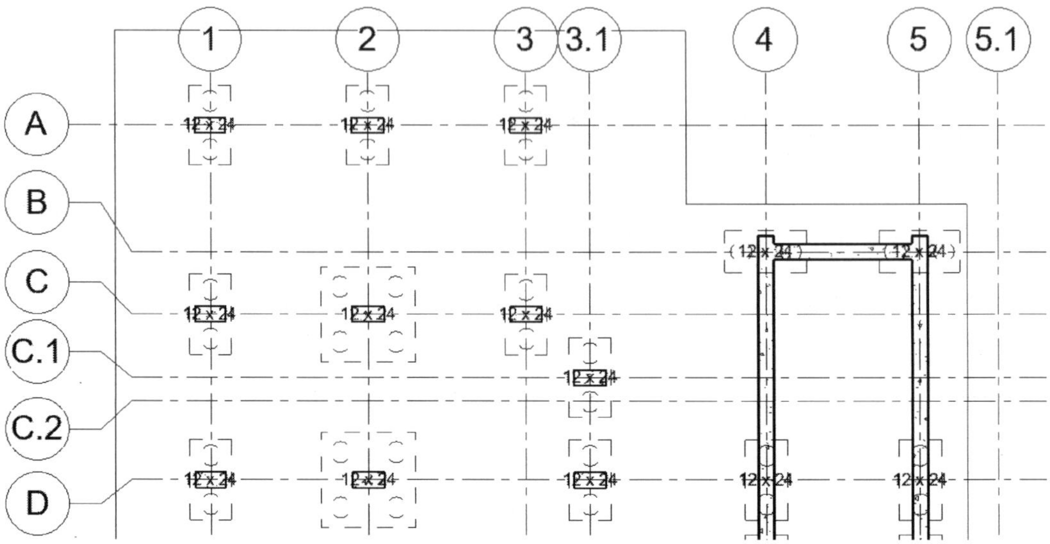

Figure 7-7 The partial view of the structural plan displaying the added structural column tag

Adding the Beam Annotations

In this section, you will tag the beams displayed in the **Second Floor** structural floor plan view.

1. Double-click on the **Second Floor** node under the **Structural Plans** head in the **Project Browser** to open the corresponding structural floor plan view.

2. Ensure that the **Annotate** tab is chosen. Choose the **Beam Annotations** tool from the **Tag** panel; the **Beam Annotations** dialog box is displayed.

3. In the **Placement** area of this dialog box, ensure that the **All beams in current plan view** radio button is selected and then select the **Remove existing beam tags and spot elevations** check box.

4. In the **Annotation location and type** area of this dialog box, ensure that the **Level beams in plan** tab is chosen.

5. In this tab, choose the Browse button located at the top middle of the image of the beam; the **Select Annotation Type** dialog box is displayed.

6. In the **Select Element to Place** area of the **Select Annotation Type** dialog box, ensure that the **Structural Framing Tag** radio button is selected. Then, in the **Structural Framing Tag** area, click in the **Type** drop-down list and then select the framing tag option from the list displayed.

   For Imperial **Structural Framing Tag : Boxed**
   For Metric **M_Structural Framing Tag : Boxed**

7. Choose the **OK** button; the **Select Annotation Type** dialog box is closed.

8. In the **Level beams in plan** tab of the **Beam Annotations** dialog box, choose the Browse button located at the bottom middle of the image of the beam; the **Select Annotation Type** dialog box is displayed.

9. In the **Select Element to Place** area of the **Select Annotation Type** dialog box, ensure that the **Spot Elevation** radio button is selected.

10. In the **Spot Elevation** area of this dialog box, select the **Beam Elevation (Relative)** option from the **Type** drop-down list and then select the **Entry** option from the **Relative Base** drop-down list.

11. Select the **Top & Bottom Elevations** option from the **Display Elevation** drop-down list in the **Spot Elevation** area.

12. Choose the **OK** button; the **Select Annotation Type** dialog box is closed. Next, choose the **OK** button in the **Beam Annotations** dialog box; the **Beam Annotations** dialog box is closed and the beam annotations are added to all the beams placed in the view, as shown in Figure 7-8.

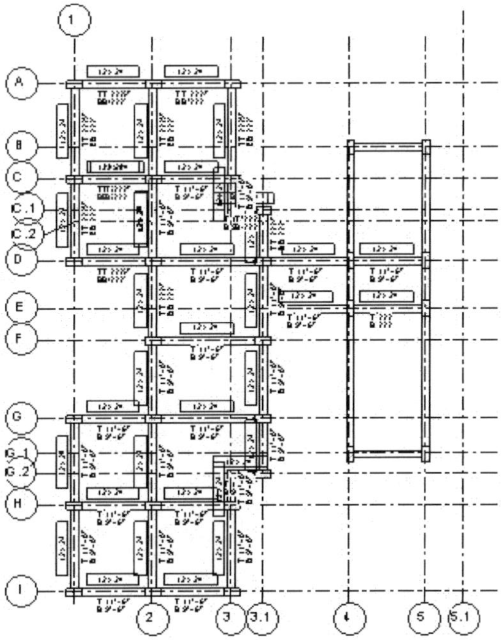

Figure 7-8 *The beam annotation displayed in the* ***Second Floor*** *structural plan view*

Saving the Project

In this section, you will save the project file by using the **Save As** tool.

1. Choose **Save As > Project** from the **File** menu; the **Save As** dialog box is displayed.

2. In this dialog box, browse to the *C:\rst_2020* location and create a folder with name **c07_rst_2020_tut**. Next, open the created folder and enter **c07_Commercial-Complex_tut1** for Imperial and **M_c07_Commercial-Complex_tut1** for Metric unit system in the **File name** edit box.

3. Now, choose the **Save** button; the **Save As** dialog box closes and the project file is saved.

Closing the Project

1. Choose the **Close** option from the **File** menu.

   The file is closed. This completes Tutorial 1 of Chapter 7.

EXERCISE

Exercise 1 Academic Institution

In this exercise, you will add dimensions to the grid lines and add tags to columns and foundations at the **Entry** structural floor plan view of the structural model created in the Exercise 1 of Chapter 6. Further, you will add Beam Annotations to the **Roof** structural plan view. Refer to Figures 7-9 and 7-10. **(Expected time: 1 hr)**

1. Use the **Entry** structural plan view to dimension the grids.
2. Use the **Roof** structural plan view to add beam annotations.
3. File name to be assigned:
   For Imperial *c07_Academic-Institution_exer1.rvt.*
   For Metric *M_c07_Academic-Institution_exer1.rvt.*

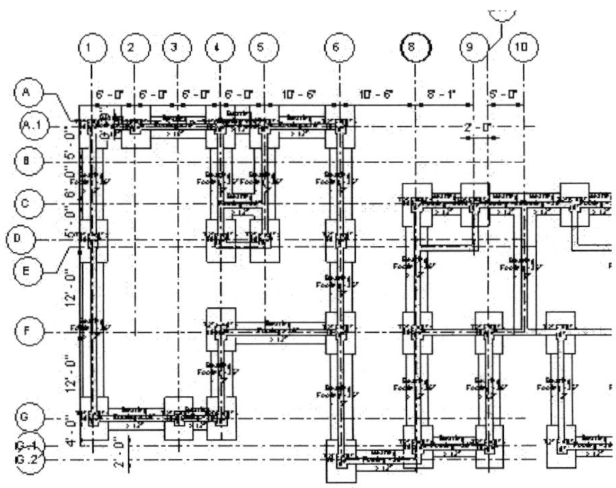

***Figure 7-9** The **Entry** structural plan view displaying the grid dimensions and the tags added to the foundations and columns*

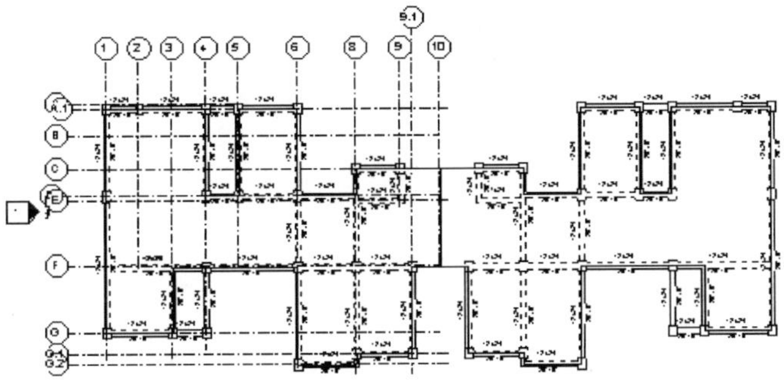

Figure 7-10 *The* ***Roof*** *structural plan view displaying the Beam Annotations*

This page is intentionally left blank

Chapter 8

Standard Views, Details, and Schedules

Learning Objectives

After completing this chapter, you will be able to:

- *Create Elevation Views*
- *Create Section Views*
- *Create Duplicate Views*
- *Generate Graphical Column Schedule*
- *Create Schedule*
- *Generate Material Takeoff Schedule*

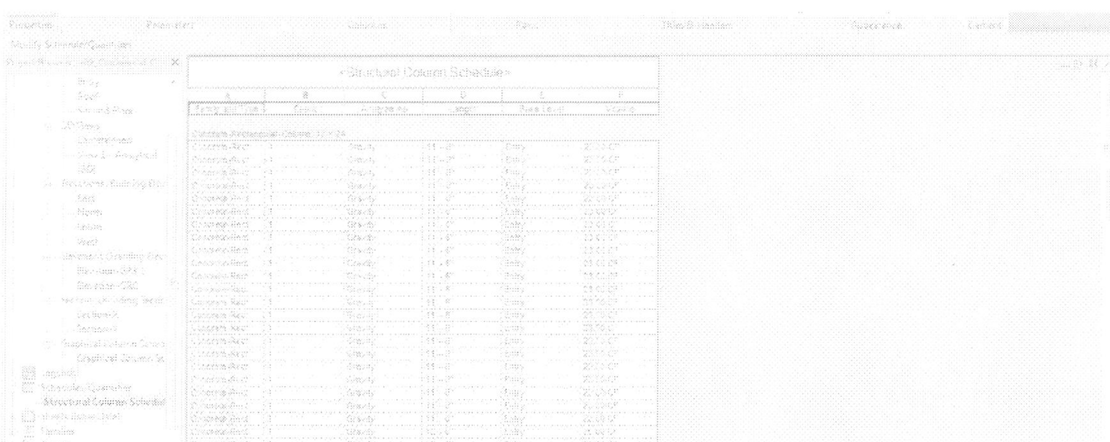

In the previous chapters, you learned to create and document a structural model. In this chapter, you will learn to generate standard views and details as well as schedule reports in a structural model.

In Revit, you can generate sections and elevations very easily. The views generated from sections and elevations will help you prepare the detail drafting of a structural model. Also, you can use these views to add various structural elements to a structural model. For example, you can use section views to add reinforcement details to the structural beam, columns, or floors. Similar to the section and elevation views, you can create drafting and detail views for the structural model.

After creating a structural model and generating the required views, the next important task is to generate the reports related to the structural model. These reports help in execution of the construction of the building on site. These reports are in the form of graphical column schedules, general schedules for estimating the quantities of various structural elements, schedules for Material Takeoffs, and others.

ELEVATION VIEWS

Elevation view is the most commonly used orthographic projection for conveying the appearance of a building model from its exterior or interior. In Revit, elevation views are part of the default template file that you can use to create your project. When you open the default template file to create a project, four elevation views are displayed: North, South, East, and West. However, you can create more exterior and interior elevation views of your structural model using various tools available in Revit.

In Revit, you can create two types of elevation views: building and framing. The building elevation views are similar to the sectional views in a model. The building elevations are capable of generating four directional views: North, South, East, and West. These views are located under the **Elevations (Building Elevation)** head in the **Project Browser**. The Framing elevation views are used to place vertical bracing and moment bracing members in a structural model. The framing elevation views are associated to a grid and therefore, it is necessary to have a grid in your view to create a framing elevation view. When you create framing elevation views in a project, the views created are located under the **Elevations (Framing Elevation)** head in the **Project Browser**. In Revit, you can create building and framing elevation views by using the **Elevation** and **Framing Elevation** tools respectively, located in the **View** tab. The methods of creating building elevation views in a structural model are discussed next.

SECTION VIEWS

Section views are generated by cutting sections through a building model. These views are created to display the wall elevations, floor heights, cross-sectional details of structural elements, and special vertical features of a project. They are also useful in creating and editing reinforcements for various structural elements in a structural model. For example, to emphasize the structural details of the central atrium of an office building model, you may need to show a section through the central atrium. Revit enables you to create the section with relative ease. You can also modify the sectional view to create a section displaying other details.

CALLOUT VIEWS

A callout view is an enlarged view of a part of a building model which requires more detailing. Creating callout views is a common practice among engineers as callout views help them look the project model more precisely and with a higher level of detail. For example, in a structural model, a callout view can be used to show the details of a connection of a beam and column in the plan or an elevation view. In a project, you can create a callout view for a plan view, section view, or elevation view. In these views, the callout tag added to a view will be linked to the callout view. The view in which the callout tag is added is called the parent view of the callout view. If the parent view is deleted, the callout will also be deleted. A callout tag, as shown in Figure 8-1, is an annotation element that represents the location of the callout in the plan, elevation, or section. The callout tag consists of the following parts: Callout bubble, Callout head, Leader line, and Reference labels. The callout bubble is the line drawn around that part of the structural model which you need to enlarge and view in the callout view. The callout head is a symbol that represents the callout view. When a callout view is placed on a sheet, the callout head displays a detail number and a sheet number by default. The line that connects the callout head to the callout bubble is called the leader line. The reference label in a callout tag is used for the callout views that are referenced.

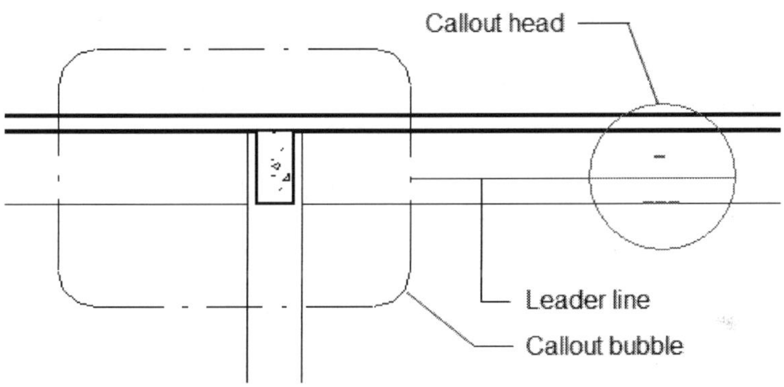

Figure 8-1 *The display of a callout tag*

DRAFTING DETAILS

Drafted details are created when you want to access the details that are not referenced to the existing project views. These details are not linked to a building model and therefore they do not get updated with it.

To create a drafted detail, first create a drafting view and then use the drafting tools provided in Revit to sketch the detail. You can also import in-built details from Revit's detail library and use them.

After the drafted detail has been created, this can be used as a reference detail. The method for creating a drafting view and drafting detail are discussed in the next sections.

DUPLICATE VIEWS

You can create multiple copies of a project view in Revit. These multiple views are called duplicate views. All duplicate views are dependent on the original view from which they are created. Therefore, these views are updated automatically if any changes are made in the original view. Duplicate views are useful when you want to use or place the same view on more than one sheet. These views are also useful in case of huge projects when the overall project view is too large to fit into a single sheet. In such cases, you can crop the parent view to make small segments by using the crop regions, create dependent views from them, and then place the cropped dependent views on the sheet. In Revit, you can create three types of duplicate views. The tools used for creating these duplicate views are: **Duplicate View**, **Duplicate with Detailing**, and **Duplicate as Dependent**. You can use the **Duplicate View** tool to create a view that contains only the model geometry from the current view. The **Duplicate with Detailing** tool can be used to create a view that contains view-specific elements from the current view. You can use the **Duplicate as Dependent** tool to create a view that is dependent on the current view. The method of creating duplicate view as a dependent view is discussed next.

GRAPHICAL COLUMN SCHEDULES

In a structural project, the Graphical Column Schedule (GCS) is a useful document which contains information related to the size, connection details, and reinforcement details of columns. The Graphical Column Schedule is a tabular representation of the columns in a project. You can use the information in the Graphical Column Schedule to track the structural columns used in the structural model. The columns displayed in the Graphical Column Schedule are linked to grid intersections. For example, if a column is located at the intersection of grid 1 and grid A, the Graphical Column Schedule will display the location of the column as A-1. If a column is not linked to a grid or grid intersection, Revit will link the column to the nearest grid or grid intersection. In the next section, the creation and use of Graphical Column Schedule in a project are discussed.

USING SCHEDULES IN A PROJECT

Schedule is another format for providing project information. In a building project, there may be different items that can be used a number of times at different locations. Schedules are primarily used to provide information regarding these items in a tabular format. For each item, the schedule tables provide information regarding its size, material, cost, finish, level, and so on. These schedules can then be used by associated agencies for various purposes. These schedule tables can assist the fabricating agency to manufacture all similar items. The quantity surveyors and estimators can calculate the quantity and cost of the grouped items. For example, in a building model, a particular column type may be used a number of times in the same layout plan on different floors. With the help of schedules, you can group columns in different ways based on their parameters such as location in a project, size, section, and so on. The project schedules are created by extracting information from a building model. In a project, all elements are parametric. This means that they have several properties associated with them. When you use these elements, the associated information is automatically added to the building model. This information can then be extracted for creating schedules.

GENERATING THE MATERIAL TAKEOFF SCHEDULE

Ribbon: View > Create > Schedules drop-down > Material Takeoff

Material takeoff schedules display the information of the material used to create structural elements in a model. This schedule helps the quantity surveyor in estimating the materials used to fabricate the structural elements in the site.

The method of generating the material takeoff schedule is the same as that of generating the quantity schedule. To generate a material takeoff schedule, choose the **Material Takeoff** tool from **View > Create > Schedules** drop-down; the **New Material Takeoff** dialog box will be displayed as shown in Figure 8-2.

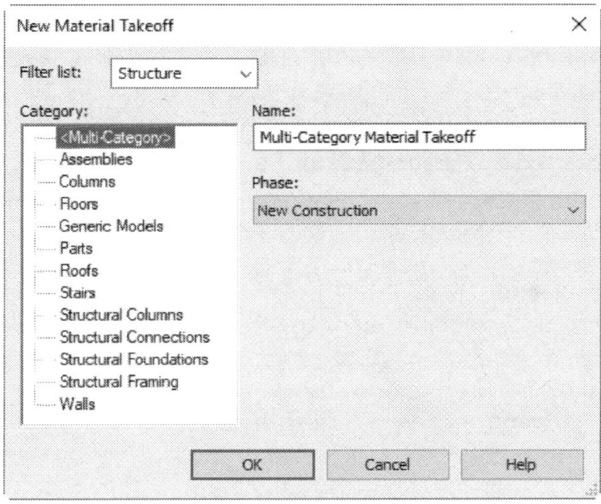

Figure 8-2 *The* ***New Material Takeoff*** *dialog box*

TUTORIAL

Tutorial 1 Commercial Complex

In this tutorial, you will create a section view, an elevation view, and a quantity schedule for the project created in Tutorial 1 of Chapter 7. **(Expected time: 1 hr 15 min)**

The following steps are required to complete this tutorial:

a. Open the *c07_Commercial-Complex_tut1.rvt* project file.
b. Hide annotations and dimensions.
c. Create section views.
d. Create the framing elevation view.
e. Generate Graphical Column Schedule.
f. Create quantity schedule for columns.
g. Save the project.
h. Close the project.

Opening the Project File

In this section, you will open the project file created in Tutorial 1 of Chapter 7.

1. To open the project file, choose **Open > Project** from the **File** menu; the **Open** dialog box is displayed.

2. In this dialog box, browse to *C:\rst_2020\c07_rst_2020_tut* folder location and then choose the *c07_Commercial-Complex_tut1.rvt* for Imperial, and for Metric choose *M_c07_Commercial-Complex_tut1.rvt* project file. You can also download this file from *https://www.cadsofttech.com*. The path of the file is as follows: *Textbooks > Civil/GIS > Revit Structure > Revit Structure 2020 for Novices*.

3. Next, choose the **Open** button from the **Open** dialog box; the selected project file opens in the drawing window.

Hiding Annotations and Dimensions

In this section, you will hide annotations and dimensions in the **Second Floor** structural plan view so that sections can be placed easily.

1. Double-click on the **Second Floor** node under the **Structural Plans** head in the **Project Browser**; the structural floor plan view at the **Second Floor** level is displayed.

2. Choose the **View** tab and then choose the **Visibility/ Graphics** tool from the **Graphics** panel; the **Visibility/Graphic Overrides for Structural Plan: Second Floor** dialog box is displayed.

3. Choose the **Annotation Categories** tab from this dialog box. In this tab, ensure that the **Structure** option is selected in the **Filter list** drop-down list. Next, in the **Visibility** column, clear the check boxes corresponding to the following categories: **Dimensions**, **Structural Column Tags**, **Structural Foundation Tags**, **Structural Framing Tags**, **Spot Coordinates**, and **Spot Elevations**.

4. Choose the **OK** button; the **Visibility/Graphic Overrides for Structural Plan: Second Floor** dialog box is closed.

Creating Section Views

In this section, you will create sectional views for the project by using the **Section** tool.

1. Choose the **Section** tool from the **Create** panel of the **View** tab; the **Modify | Section** contextual tab is displayed.

2. In the **Properties** palette, ensure that the **Section : Building Section** option is selected in the **Type Selector** drop-down list.

3. Move the cursor between the grid bubbles of grid D and grid E, and then click to specify the start point of the grid line.

4. Move the cursor horizontally across the building model till you have crossed the vertical grid 5.1. Click at the desired point, as shown in Figure 8-3, to specify the endpoint of the section line. The section view is created and the name **Section 1** is displayed in the **Sections (Building Section)** head in the **Project Browser**.

5. In the **Properties** palette, scroll the palette, if required and then click in the value field of the **View Name** parameter under the **Identity Data** area, and enter **Section-X** to replace the existing value.

6. Next, click in the value field corresponding to the **Title on Sheet** parameter under the **Identity Data** area and enter **Sectional View-Longitudinal** in it.

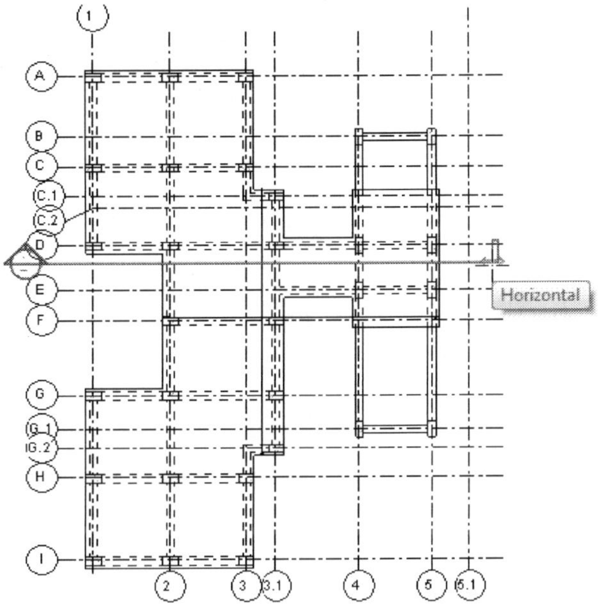

Figure 8-3 *Creating a section line for the section view*

7. In the **Properties** palette, click in the value field corresponding to the **Detail Level** parameter under the **Graphics** area and select the **Fine** option from the drop-down list displayed.

8. Next, click in the value field of the **Far Clip Offset** parameter under the **Extents** area and enter **20**'(**6096** mm) in it. Choose the **Apply** button in the **Properties** palette.

9. Now, choose the **Split Segment** tool from the **Section** panel of the **Modify | Views** contextual tab; a pencil mark appears in place of the cursor.

10. Place the cursor over the section line; the section line gets highlighted. Next, click over the section line at its intersection with grid 4. As you move down the cursor toward the right, the split section line also moves. Place the cursor along with the split cursor line at grid F and click; the section line is displayed, as shown in Figure 8-4.

    The section view is created and the name **Section-X** is displayed under the **Sections (Building Section)** head in the **Project Browser**.

11. Repeat steps 1 to 9 to create a transverse section view of the building model. Name the view as **Section-Y** and the title for the sheet as **Sectional View-Transverse** and set the **Far Clip Offset** parameter to **45**'(**13716** mm). For the location of the section line, refer to Figure 8-5.

12. Click on the '+' symbol on the left of the **Sections (Building Section)** head in the **Project Browser** to display the section names created. The **Section X** and **Section Y** is created under the **Sections(Building Section**) head.

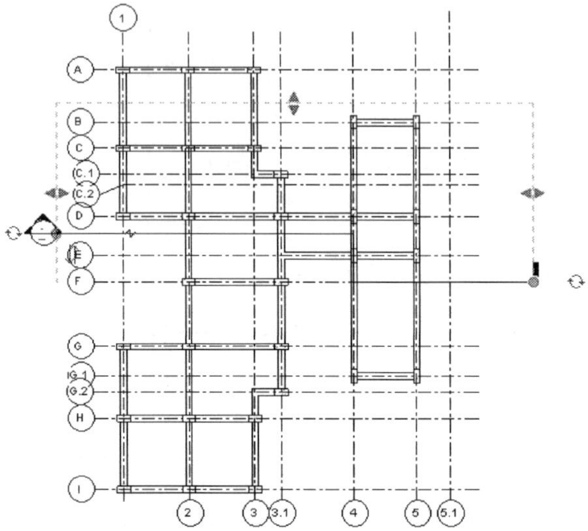

Figure 8-4 *Section line for longitudinal section after splitting the segment*

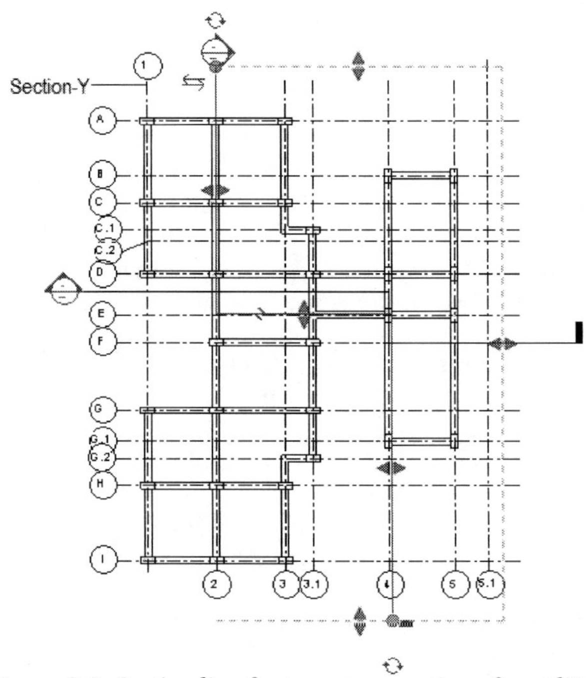

Figure 8-5 *Section line for transverse section after splitting the segment*

13. Press ESC to exit the current selection.

14. Double-click on the name **Section-X** to display the corresponding section view on the drawing window, as shown in Figure 8-6.

15. Similarly, double-click on the name **Section-Y** to display the corresponding section view in the drawing window, as shown in Figure 8-7.

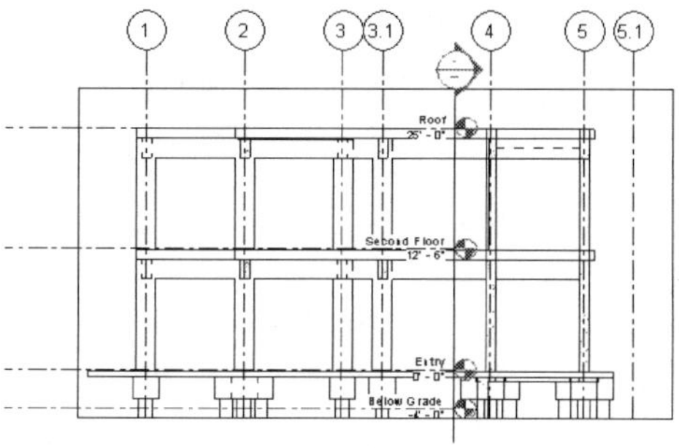

Figure 8-6 *Section view for* ***Section-X***

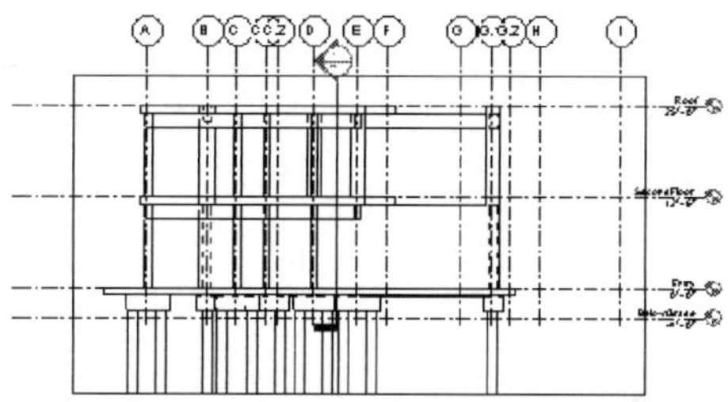

Figure 8-7 *Section view for* ***Section-Y***

Note
You can drag the drag points of the ***Crop Region*** *to customize the view of your section.*

Creating Framing Elevation Views

In this section, you will create the elevation views for the project using the **Framing Elevation** tool and the **Building Elevation** tool.

1. Double-click on the **Second Floor** node under the **Structural Plans** head of the **Project Browser**.

2. Invoke the **Framing Elevation** tool from **View > Create > Elevation** drop-down.

3. Place the cursor over grid C; it is highlighted and an elevation symbol is displayed. Note that the elevation symbol will flip above and below the grid line as you move the cursor.

4. Move the cursor slightly below grid C such that the elevation symbol is displayed below the highlighted grid line.

5. Move the cursor towards right and place it at the location, as shown in Figure 8-8, and click; the elevation tag is added. Also, an elevation view is created and added under the **Elevations (Framing Elevation)** head in the **Project Browser**.

6. Press ESC and then select the arrow head of the elevation tag just added; the **Modify | Views** contextual tab is displayed. In the **Properties** palette, click in the value field corresponding to the **View Name** parameter under **Identity Data** head and enter **Elevation-GRC** in it. Choose the **Apply** button.

7. Press ESC twice to exit the current selection.

8. Repeat steps 2 to 6 and create an elevation at grid 3.1, refer to Figure 8-9. Then, name it as **Elevation GR-3.1**.

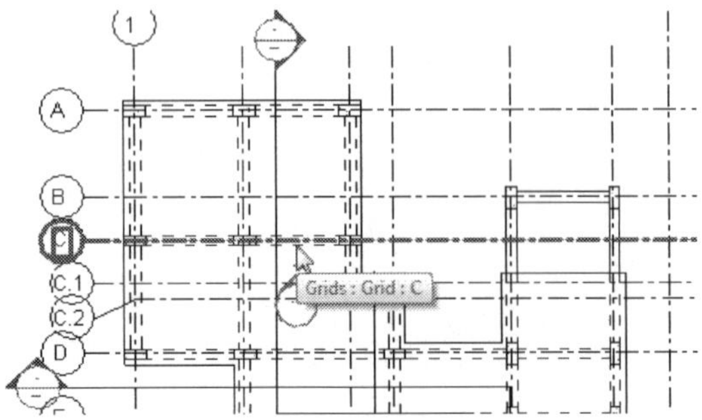

Figure 8-8 Placing the elevation tag at grid C

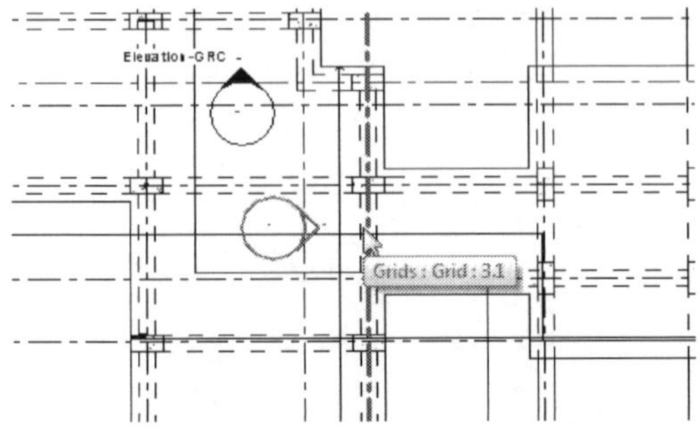

Figure 8-9 Placing the elevation tag at grid 3.1

9. Double-click on **Elevation-GRC** under the **Elevations (Framing Elevation)** node in the **Project Browser** to display the elevation view, as shown in Figure 8-10.

10. Double-click on **Elevation-GR3.1** to display its corresponding view, as shown in Figure 8-11.

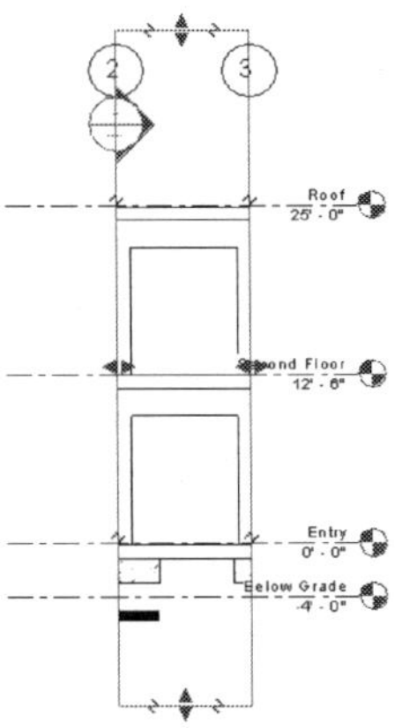

*Figure 8-10 The **Elevation-GRC** sectional view*

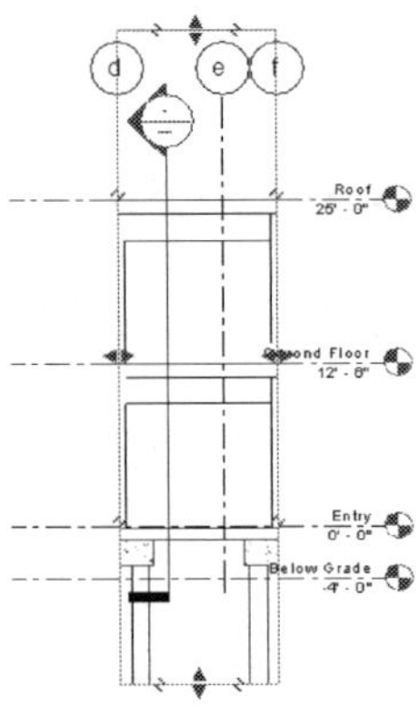

*Figure 8-11 The **Elevation-GR3.1** sectional view*

Generating the Graphical Column Schedule

In this section, you will create the graphical schedule for column and the quantity schedule for columns using the **Graphical Column Schedule** tool.

1. Invoke the **Graphical Column Schedule** tool from **View > Create > Schedule** drop-down; the graphical column schedule is displayed in the drawing window.

2. In the **Properties** palette, select the check box corresponding to the **Group Similar Locations** parameter under the **Graphics** area; the columns at the similar grid line are grouped and displayed in the schedule.

3. Now, click in the **View Scale** parameter under the **Graphics** area and then select the **1/4" = 1'-0" (1:25)** option from the drop-down list displayed. Next, choose the **Apply** button; the viewing scale of the Graphical Column Schedule is changed, as shown in Figure 8-12.

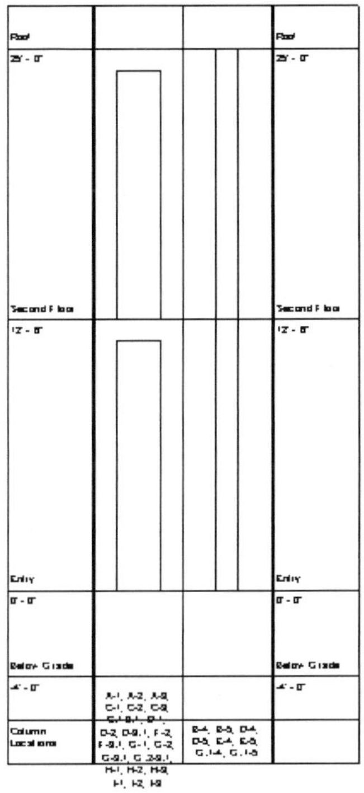

Figure 8-12 The ***Graphical Column Schedule*** *view*

Note

*As you create the schedule, the name of the schedule is displayed as **Graphical Column Schedule 1** under the **Graphical Column Schedule** head in the **Project Browser**.*

Generating a Quantity Schedule for Columns

In this section, you will create a schedule for displaying the quantity columns and their associated properties used in the project.

1. Invoke the **Schedule/Quantities** tool from **View > Create > Schedules** drop-down; the **New Schedule** dialog box is displayed.

2. In this dialog box, select the **Structural Columns** option from the **Category** list box and then choose the **OK** button; the **Schedule Properties** dialog box is displayed.

3. In the **Available fields** list box, select the **Count** option. Now, press and hold the CTRL key and select the following options from the list box: **Base Level**, **Family and Type**, **Length**, and **Volume**.

4. Next, release the CTRL key and choose the **Add parameter(s)** button; the selected options are added to the **Scheduled fields (in order)** list box.

5. Click on the **Family and Type** option in the **Scheduled fields (in order)** list box and choose the **Move parameter up** button till the option comes at the top in the list.

6. Repeat step 5 to reorder all the options in the **Scheduled fields (in order)** list box, as shown in Figure 8-13.

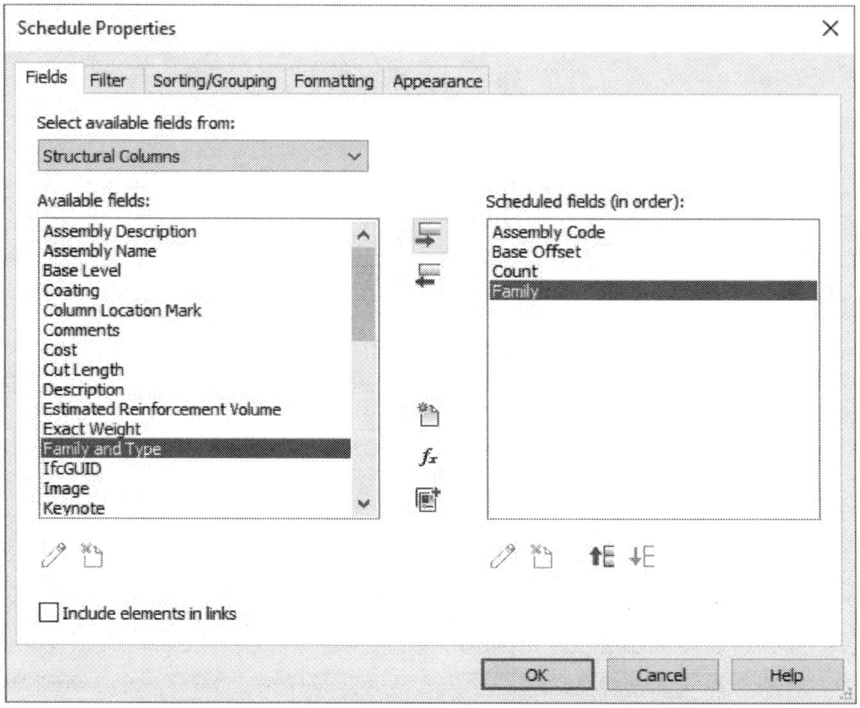

__Figure 8-13__ The __Schedule Properties__ dialog box

7. In the **Schedule Properties** dialog box, choose the **Sorting/Grouping** tab and then select the **Family and Type** option from the **Sort by** drop-down list. Ensure that the **Ascending** radio button is selected.

8. Select the **Header** and **Footer** check boxes and then ensure that the **Title, count, and totals** options are selected in the drop-down lists next to them.

9. Select the **Grand totals** check box and ensure that the **Itemize every instance** check box is selected.

10. Choose the **Formatting** tab and then in the **Fields** list box, select **Count**. Next, select the **Calculate totals** option from drop-down list below the **Show conditional format on sheets** check box.

11. Repeat step 10 and select the **Calculate totals** check box for the **Length** and **Volume** options in the **Fields** list box.

12. Choose the **OK** button; the **Structural Column Schedule** is displayed, as shown in Figure 8-14.

Note
*When you create the schedule, its name is displayed as **Structural Column Schedule** under the **Schedules/Quantities (all)** head in the **Project Browser**.*

Saving the Project

In this section, you will save the project file using the **Save As** tool.

1. To save the project, choose the **Save As > Project** option from the **File** menu; the **Save As** dialog box is displayed.

2. In this dialog box, browse to *C:\rst_2020* and create a folder with the name *c08_rst_2020_tut*. Now, open the created folder and then enter **c08_Commercial-Complex_tut1** in the **File name** edit box.

3. Choose the **Save** button; the **Save As** dialog box is closed and the project file is saved.

Closing the Project

1. To close the project, choose the **Close** option from the **File** menu.

   The file is closed. This completes Tutorial 1 of Chapter 8.

<Structural Column Schedule>				
A	B	C	D	E
Family and Type	Count	Length	Base Level	Volume
Concrete-Rectangular-Column: 12 x 24				
Concrete-Recta	1	11' - 6"	Entry	23.00 CF
Concrete-Recta	1	11' - 6"	Entry	23.00 CF
Concrete-Recta	1	11' - 6"	Entry	23.00 CF
Concrete-Recta	1	11' - 6"	Entry	23.00 CF
Concrete-Recta	1	11' - 6"	Entry	23.00 CF
Concrete-Recta	1	11' - 6"	Entry	23.00 CF
Concrete-Recta	1	11' - 6"	Entry	23.00 CF
Concrete-Recta	1	11' - 6"	Entry	23.00 CF
Concrete-Recta	1	11' - 6"	Entry	23.00 CF
Concrete-Recta	1	11' - 6"	Entry	23.00 CF
Concrete-Recta	1	12' - 6"	Entry	11.50 CF
Concrete-Recta	1	12' - 6"	Entry	11.50 CF
Concrete-Recta	1	12' - 6"	Entry	11.50 CF
Concrete-Recta	1	12' - 6"	Entry	11.50 CF
Concrete-Recta	1	11' - 6"	Entry	23.00 CF
Concrete-Recta	1	11' - 6"	Entry	23.00 CF
Concrete-Recta	1	11' - 6"	Entry	23.00 CF
Concrete-Recta	1	11' - 6"	Entry	23.00 CF
Concrete-Recta	1	11' - 6"	Entry	23.00 CF
Concrete-Recta	1	11' - 6"	Entry	23.00 CF
Concrete-Recta	1	11' - 6"	Entry	23.00 CF
Concrete-Recta	1	11' - 6"	Entry	23.00 CF
Concrete-Recta	1	11' - 6"	Entry	23.00 CF
Concrete-Recta	1	11' - 6"	Entry	23.00 CF
Concrete-Recta	1	11' - 6"	Entry	23.00 CF
Concrete-Recta	1	11' - 6"	Entry	23.00 CF
Concrete-Recta	1	12' - 6"	Entry	25.00 CF
Concrete-Recta	1	11' - 6"	Second Floor	23.00 CF
Concrete-Recta	1	11' - 6"	Second Floor	23.00 CF
Concrete-Recta	1	11' - 6"	Second Floor	23.00 CF
Concrete-Recta	1	11' - 6"	Second Floor	23.00 CF
Concrete-Recta	1	11' - 6"	Second Floor	23.00 CF
Concrete-Recta	1	11' - 6"	Second Floor	23.00 CF
Concrete-Recta	1	11' - 6"	Second Floor	23.00 CF
Concrete-Recta	1	11' - 6"	Second Floor	23.00 CF
Concrete-Recta	1	11' - 6"	Second Floor	23.00 CF
Concrete-Recta	1	11' - 6"	Second Floor	23.00 CF
Concrete-Recta	1	12' - 6"	Second Floor	23.00 CF
Concrete-Recta	1	12' - 6"	Second Floor	23.00 CF
Concrete-Recta	1	12' - 6"	Second Floor	23.00 CF
Concrete-Recta	1	12' - 6"	Second Floor	23.00 CF
Concrete-Recta	1	11' - 6"	Second Floor	23.00 CF
Concrete-Recta	1	11' - 6"	Second Floor	23.00 CF
Concrete-Recta	1	11' - 6"	Second Floor	23.00 CF
Concrete-Recta	1	11' - 6"	Second Floor	23.00 CF
Concrete-Recta	1	11' - 6"	Second Floor	23.00 CF
Concrete-Recta	1	11' - 6"	Second Floor	23.00 CF
Concrete-Recta	1	11' - 6"	Second Floor	23.00 CF
Concrete-Recta	1	11' - 6"	Second Floor	23.00 CF
Concrete-Recta	1	11' - 6"	Second Floor	23.00 CF
Concrete-Recta	1	11' - 6"	Second Floor	23.00 CF
Concrete-Recta	1	11' - 6"	Second Floor	23.00 CF
Concrete-Recta	1	11' - 6"	Second Floor	23.00 CF
Concrete-Recta	1	12' - 6"	Second Floor	25.00 CF
Concrete-Recta	1	12' - 6"	Second Floor	25.00 CF
Concrete-Recta	1	12' - 6"	Second Floor	25.00 CF
Concrete-Recta	1	12' - 6"	Second Floor	25.00 CF
Concrete-Rectang	60	706' - 0"		1350.00 CF
Grand total: 60	60	706' - 0"		1350.00 CF

Figure 8-14 *The schedule view for columns*

EXERCISE

Exercise 1 Academic Institution

In this exercise, you will create the section and framing elevation views and a callout view for the *Academic Institution* project created in Exercise 1 of Chapter 7. For the locations of section line, elevation tag, and callout, refer to Figure 8-15. Also, you will add a callout.

(Expected time: 30 min)

1. Use the **Roof** structural plan view to create the section, elevation, and callout views.
2. File to be referred: *c07_Academic-Institution_exer1.rvt.*
3. File name to be assigned: *c08_Academic-Institution_exer1.rvt.*

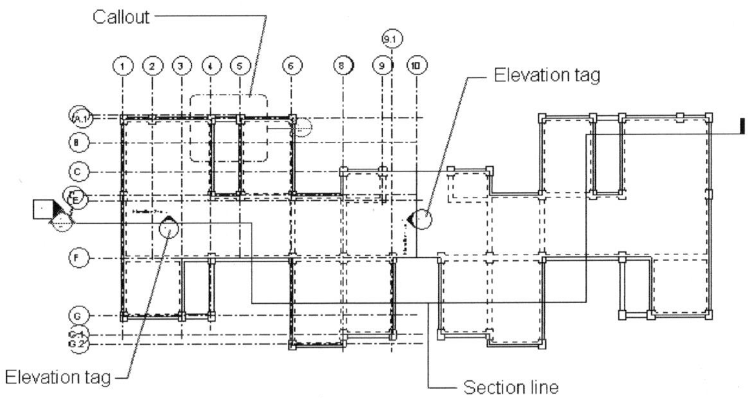

Figure 8-15 *The **Roof** level structural plan view displaying the section line, elevation tag, and callout*

This page is intentionally left blank

Chapter 9

3D Views, Sheets, Analysis, and Reinforcements

Learning Objectives

After completing this chapter, you will be able to:

- *Create 3D views*
- *Generate shadows*
- *Create sheets*
- *Use analytical models*
- *Add reinforcements*
- *Link building models and share coordinates*
- *Create massing geometry*

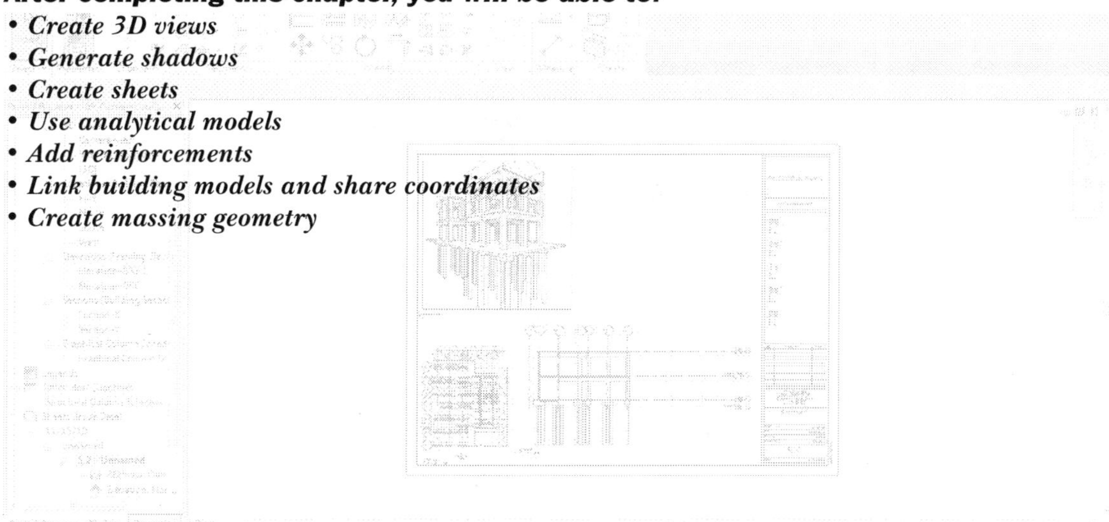

INTRODUCTION

In this chapter, you will learn how to generate three-dimensional views and 3D section views, add sheets, generate shadows, work with analytical models, and prepare the analytical model for analysis. You will also learn about adding reinforcements to concrete structural elements in a project.

THREE-DIMENSIONAL (3D) VIEWS

In Autodesk Revit, you can create and display two types of three-dimensional (3D) views, Orthographic 3D View, and Perspective View.

Orthographic 3D View: An orthographic 3D view is a 3D view where all components are of same size regardless of how far they are from the camera. This is a default 3D view. Once this view has been displayed, you can then use various tools to modify its properties.

Perspective View: A perspective view is a real eye view. It can be created by placing a camera at eye elevation and specifying its target position.

Creating Orthographic 3D Views

Ribbon:	View > Create > 3D View drop-down > Default 3D View

An orthographic 3D view is the view in which all elements are displayed in their actual size, irrespective of their distance from the source. In Autodesk Revit, the default 3D view is an orthographic view. To view the orthographic 3D view of a building model, invoke the **Default 3D View** tool from the **Create** panel; the current view will change into the default 3D view. If you are creating the orthographic 3D view for the first time in a project, the camera is automatically placed at a default position (southeast corner) and the corresponding view is displayed in the drawing window. The default view is named as **{3D}** and is added under the **3D Views** head in the **Project Browser**. Figure 9-1 shows an example of the 3D view of the structural model of a building displayed using the **Default 3D View** tool.

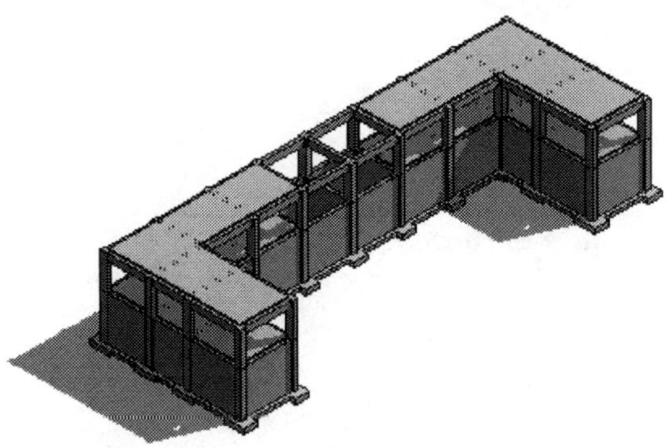

Figure 9-1 *A three-dimensional view of the structural model of a building*

Creating Perspective Views

Ribbon: View > Create > 3D View drop-down > Camera

Perspective views provide a realistic view of the exterior or interior space of a building model. Whenever you see an exterior or an interior view of a building with naked eyes, it appears as the perspective view with the wall edges of the building converging at imaginary points. You can easily create the perspective views of a building model by defining the camera location and the target point. In most building projects, you will find it easier to use a floor plan view to define these points (camera location and camera target). You can then modify the view parameters to achieve the desired result.

GENERATING SHADOWS

One of the powerful features provided in Autodesk Revit is its ability to generate shadows based on location of the Sun with respect to the building. Shadows can be generated in a plan, elevation, section, or 3D view. To use this feature, open the desired view and click on the inclined arrow symbol (**Graphic Display Options**) in the **Graphics** panel of the **View** tab; the **Graphic Display Options** dialog box will be displayed, as shown in Figure 9-2. In this dialog box, you can specify the settings to enhance the visual output of the project view. You can use the **Style** drop-down list to select an option to specify pre-defined visual settings in the project view. In the **Style** drop-down list, if you select the **Shaded** or **Realistic** option, the **Show Edges** check box will be activated. On selecting this check box, the edges of the model will be displayed in the pre-defined visual settings. You can enhance the display of your project by using the **Smooth lines with anti-aliasing** check box. On selecting this check box, the lines and edges in the view will be displayed more smoothly. In the **Model Display** area, you can drag the **Transparency** slider to set the transparency of the entire project view. Also, you can use the **Silhouettes** drop-down list to select an option to create silhouette in the model.

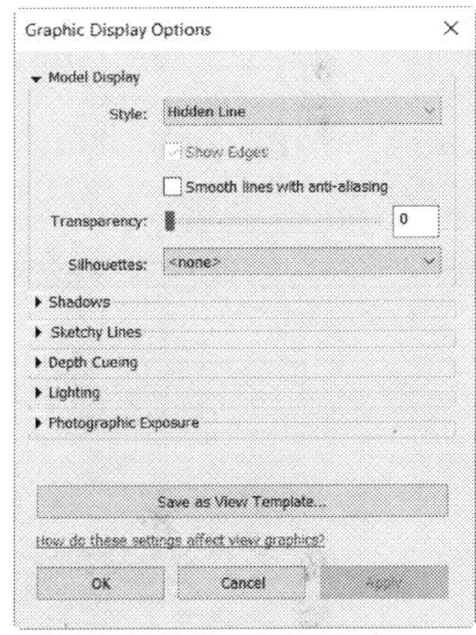

Figure 9-2 *The* ***Graphic Display Options*** *dialog box*

SHEETS

Ribbon: View > Sheet Composition > Sheet

In Autodesk Revit, a sheet or a drawing sheet is a document set that is used for the final working drawings in a site. A drawing sheet contains sheet views that consist of multiple drawing views or schedules added in a project. Sheets are defined by a border and a title block. To create a drawing sheet, you first need to add a sheet view to the project and then add the required project views to the added sheet views.

ANALYTICAL MODELS

Autodesk Revit is a multi-material (steel, precast, cast-in-place concrete, masonry, wood, and so on) modeling software for structural design, analysis, and documentation. A physical model is made up of beams, walls, columns, slabs, and footings. An analytical model is a simplified representation of the physical model in which all members are connected to each other continuously. An analytical model contains loads, load case and load combinations specific to the project being analyzed. An analytical model also includes material properties and release conditions for structural analysis.

A single structural model integrates the physical model (for layout, coordination and documentation) with an independent editable analytical model. This integration enables you to link the structural model bidirectionally to leading analysis applications such as ADAPT-Builder concrete design, RISA, Fastrak Building Designer, Autodesk Robot Structural Analysis, and other application partners. To know more about various software packages that partner with Autodesk Revit, visit *www.autodesk.com.*

Figures 9-3 and 9-4 show the 3D views of a structural model as a physical model and an analytical model, respectively. In Figure 9-4, each element in the project consists of a set of structural member analytical models. In a project, the following structural elements have structural member analytical models: Structural Columns, Structural Framing elements (such as beams and braces), Structural Floors, and Structural Walls.

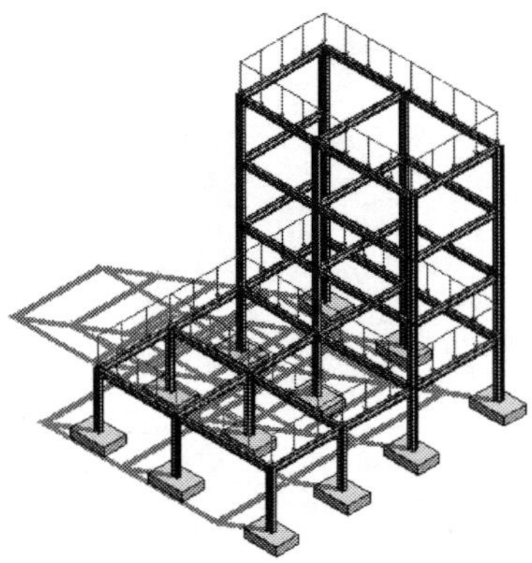

Figure 9-3 *The physical model view of a project*

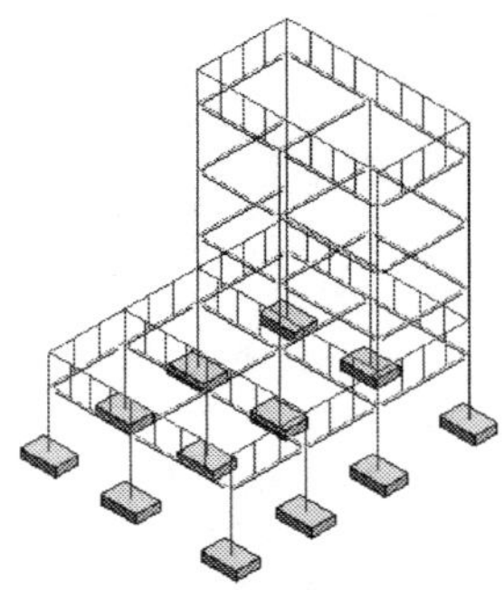

Figure 9-4 *The analytical model view of a project*

Setting Boundary Conditions

Ribbon: Analyze > Boundary Conditions > Boundary Conditions

In an analytical model, boundary conditions are model elements that define the support conditions of a structural element (beams, columns, or floors). These elements are used to add the assumed support conditions to structural elements so that they can be communicated easily to the analysis software to which a model will be exported for analysis.

To specify a boundary condition for a structural model, open an analytical structural plan view and then choose the **Boundary Conditions** tool from the **Analytical Model** panel of the **Analyze** tab; the **Modify | Place Boundary Conditions** contextual tab will be displayed, as shown in Figure 9-5. The **Boundary Conditions** panel of this tab is used to add different types of boundary conditions.

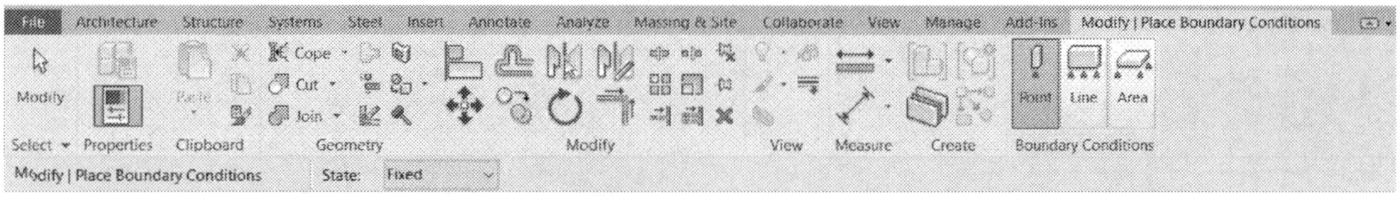

*Figure 9-5 Various options in the **Modify / Place Boundary Condition**s tab*

ADDING REINFORCEMENTS

Reinforcements are added to concrete structural members based on the design requirement. In Revit, you can add reinforcements to concrete structural members such as floors, beams, columns, and slabs by using the **Area** tool, **Path** tool, and **Fabric Area** tool.

LINKING BUILDING MODELS AND SHARING COORDINATES

Large projects may require to develop independent building models and later combine these models into a comprehensive project. For example, in a large educational campus project, you need to independently develop the building models of the administrative building, academic complex, faculty residences, student accommodation, and so on, and later add them into a single host project file. You can do so by using the linking and sharing coordinate tools of Autodesk Revit that enable you to link models and share their coordinates within a single host project file. These tools also enable you to work on building models in other project files. When you link projects, you need to share the coordinates of the host model with the linked model so that the linked files retain their positions. To do so, invoke the **Acquire Coordinates** tool from **Manage > Project Location > Coordinates** drop-down and then select the linked project in your drawing; the origin of the shared coordinates for the linked project becomes the origin for the shared coordinates of the host project. Revit provides flexibility and easy management of the linked models by enhancing the linking of the models and organizing the linked files in the **Project Browser**. You can easily access the linked files, the nested link files, and the Link Manager from the **Project Browser**. The nested Revit links are also listed under the **Revit Links** head with the host link in the **Project Browser**.

Linking or Importing Files

In Revit, you can link or import the CAD files to the current project. If a CAD file is linked to a project, all changes made in it will simultaneously be updated in the project. However, if you import a file in a project, it will remain unchanged, irrespective of the changes made in the file linked to the project.

INTRODUCTION TO MASSING

In the earlier chapters, you learned how to use different building elements and components to create a building model. These building elements are parametrically associated and enable you to generate a building model based on specific design requirements such as wall types, door width, window height, and so on. Each of the elements must be assigned specific properties to achieve the desired element parameters, thereby making a building model accurate. Needless to say, this is a fairly time-consuming procedure. Autodesk Revit provides you with an alternative and much easier method to create a building model, known as massing.

TUTORIAL

Tutorial 1 Commercial Complex

In this tutorial, you will create a camera view in the **Entry** structural plan view of the *Commercial Complex* project for Imperial or *M_Commercial Complex* project for Metric created in Tutorial 1 of Chapter 8. Also, you will generate shadows, create a sheet, and add hosted area loads to the structural elements in the project. **(Expected time: 45 min)**

The following steps are required to complete this tutorial:

a. Open the existing project file.
   For Imperial *c08_Commercial-Complex_tut1.rvt*
   For Metric *M_c08_Commercial-Complex_tut1.rvt*
b. Create a camera view.
c. Generate shadows.
d. Create a sheet and add project view to it.
e. Add hosted area loads.
f. Save the project by using the **Save As** tool.
g. Close the project by using the **Close** tool.

Opening the Existing Project

In this section, you will open the project file created in Tutorial 1 of Chapter 8.

1. Choose **Open > Project** from the **File** menu; the **Open** dialog box is displayed.

2. In this dialog box, browse to *C:\rst_2020\c08_rst_2020_tut* location and choose the project file *c08_Commercial-Complex_tut1.rvt* for Imperial or *M_c08_Commercial-Complex_tut1.rvt* for Metric. You can also download this file from *https://www.cadsofttech.com*. The path of the file is as follows: *Textbooks > Civil/GIS > Revit Structure > Revit Structure 2020 for Novices*.

3. Choose the **Open** button from the **Open** dialog box; the selected project file with a schedule opens in the drawing window.

Creating a Camera View

In this section, you will create a camera view for the *Commercial Complex* project. The camera view will be created at a height of **5'6"** for Imperial or **1676** mm for Metric from the **Entry** level of the project.

1. Double-click on the **Entry** node under the **Structural Plans** head in the **Project Browser**.

2. Invoke the **Camera** tool from **View > Create > 3D View** drop-down; the options for inserting the camera are displayed in the Options Bar.

3. In the Options Bar, ensure that the **Perspective** check box is selected. Enter the value **5'6"** for Imperial or **1676** mm for Metric in the **Offset** edit box and then press ENTER.

4. Move the cursor down and click at a position between the **West** and **South** elevation arrows to specify the camera location. For the location of the camera, refer to Figure 9-6.

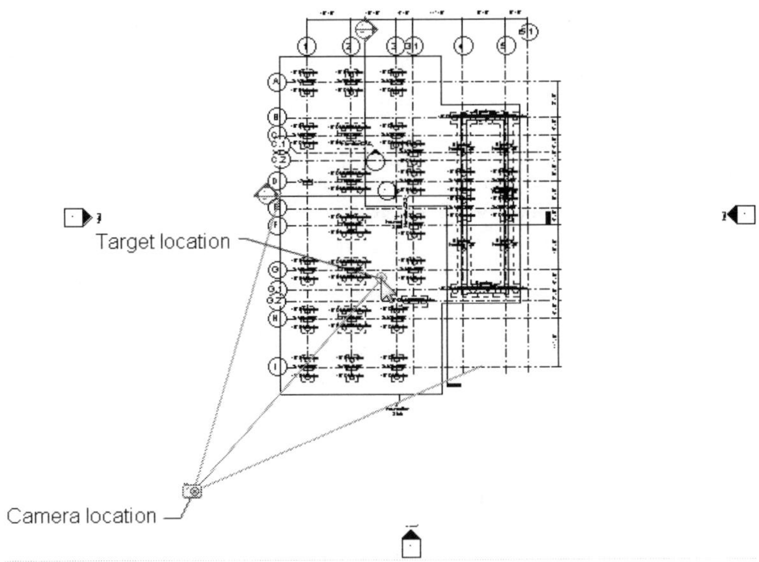

Figure 9-6 *Structural plan view showing the location of the camera and its target*

5. Move the cursor between the grid G, G.1 and 2, 3 location and press ENTER, refer to Figure 9-6; the project view gets changed and a view similar to the view shown in Figure 9-7 is displayed.

6. In this view, ensure that the crop region of the 3D view is selected and the drag controls (blue dots at the midpoint of the four sides of the rectangle) are displayed.

7. Use drag controls to increase the extent of the view so that the entire structural model is visible inside the crop-region.

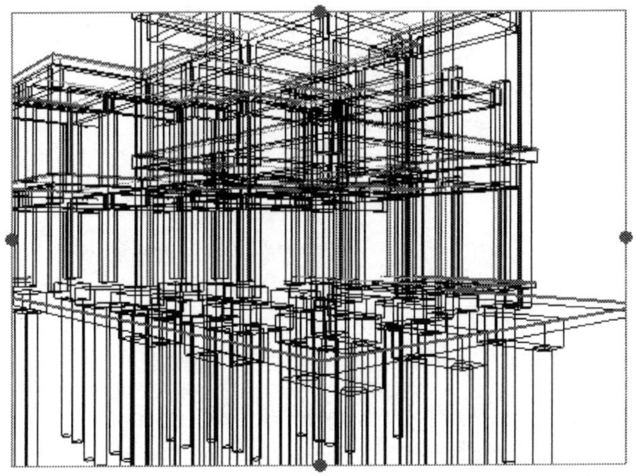

Figure 9-7 *Camera view displaying the structural model in wireframe*

8. Ensure that the **Crop Region Visible** is selected in the **Properties** palette, click in the value field of the **View Name** parameter and enter **Camera-view** in it to replace the existing value. Similarly, click in the value field of the **Title on Sheet** parameter and enter **Perspective-View** in it, and choose the **Apply** button.

Generating Shadows

In this section, you will adjust the Sun settings and generate shadows of the structural model displayed in the camera view.

1. Ensure that the **Camera-view** view is displayed in the **3D Views** head from the **Project Browser**. And then, choose the **Visual Style: Wireframe** button from the **View Control Bar**; a flyout is displayed. Next, choose the **Hidden Line** option from the flyout displayed; the hidden line view of the structural model is displayed, as shown in Figure 9-8.

2. Choose the **Sun Path Off** button from the **View Control Bar**; a flyout is displayed.

3. In the displayed flyout, choose the **Sun Settings** options; the **Sun Settings** dialog box is displayed.

4. In this dialog box, select the **Still** radio button in the **Solar Study** area. In the **Settings** area, ensure that the **Location** parameter is set to **Boston, MA**. Choose the **Disable Updater** button from the **Autodesk Revit 2020** message box if displayed. The **Sun Settings** dialog box is displayed again.

5. Make sure that the **Ground Plane at Level** check box is selected in the **Sun Settings** dialog box. Next, select the **Entry** option from the drop-down list displayed below this check box.

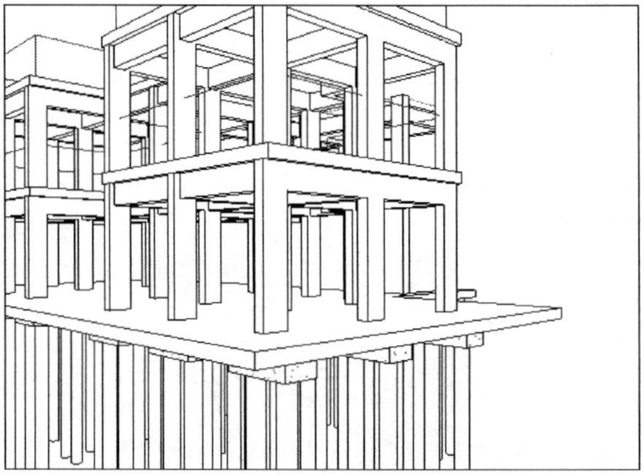

Figure 9-8 *Camera view displaying the structural model in the **Hidden Line** visual style*

6. Choose the **Apply** and then the **OK** button; the **Sun Settings** dialog box is closed.

7. In the **View Control Bar**, choose the **Shadows Off** button; a shadow is generated in the structural model. Figure 9-9 shows the camera view of the structural model displaying shadows due to the Sun settings.

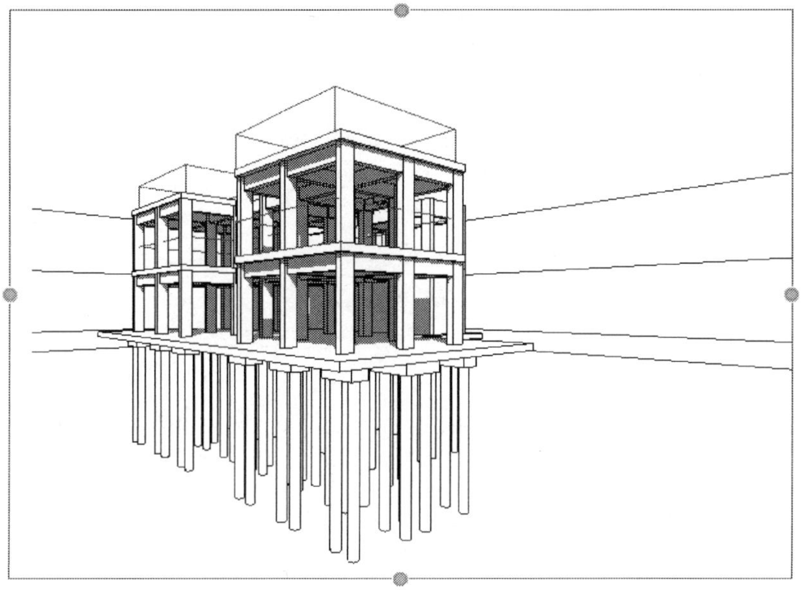

Figure 9-9 Camera view displaying the structural model with shadows

Creating a Sheet

In this section, you will create a sheet in the project.

1. Choose the **Sheet** tool from the **Sheet Composition** panel in the **View** tab; the **New Sheet** dialog box is displayed.

2. In this dialog box, choose the **Load** button; the **Load Family** dialog box is displayed.

3. In this dialog box, select the specified title block **C 17 x 22 Horizontal** from the **US Imperial > Titleblocks** folder for Imperial unit system and load **A0 metric** from the **US Metric > Titleblocks** folder for Metric unit system. Choose the **Open** button from the **Load Family** dialog box. On doing so, the specified title block is added to the list in the **Select titleblocks** region of the **New Sheet** dialog box.

4. Choose the **OK** button to create the sheet view by using the loaded title block. The added sheet is now displayed in the drawing window.

Adding Project Views to the Sheet

In this section, you need to add the specified project views to the sheet by dragging their name from the **Project Browser**. Further, based on the sheet layout, you need to place the project views at their designated place in the sheet.

1. In the **Project Browser**, click on the **Entry** node under the **Structural Plans**. Next, in the **Properties** palette, click in the value field corresponding to the **View Scale** parameter and select the **1/16" = 1'-0"** option for Imperial and **1:200** option for Metric from the drop-down list displayed.

2. Now, press and hold the left mouse button on the **Entry** node in the **Project Browser** and then drag the view into the drawing sheet. Release the left mouse button when the project view appears as a rectangle in the sheet.

3. Move the cursor to the lower left area of the title block such that the corner of the rectangle get close to the lower left corner of the drawing sheet. Next, click to place the view; the **Entry** structural plan view is added to the sheet and appears enclosed in a rectangle, as shown in Figure 9-10.

4. Next, ensure that the viewport of the **Entry** structural plan view is selected in the sheet. In the **Properties** palette, ensure that the check box corresponding to the **Crop Region Visible** parameter is cleared.

5. Choose the **Edit Type** button in the **Properties** palette; the **Type Properties** dialog box is displayed. In the **Type Properties** dialog box, select the **Title Only** option from the **Type** drop-down list and ensure that the check box corresponding to the **Show Extension Line** parameter is cleared and then choose the **OK** button.

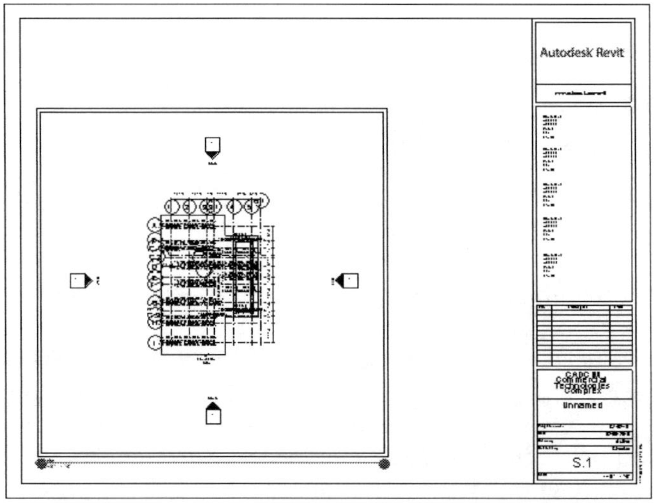

***Figure 9-10** The location of the **Entry** structural plan view in the sheet*

6. Right-click in the drawing and then choose the **Activate View** option from the shortcut menu displayed.

7. Now, in the project view, click on the **North** elevation arrow head and right-click; a shortcut menu is displayed. Choose **Hide in View > Category** from it.

8. Again, right-click and then choose the **Pan Active View** option from the shortcut menu displayed; a move icon is displayed with the cursor.

9. In the sheet, press and drag the **Entry** structural plan view and then place it at the location, shown in Figure 9-11.

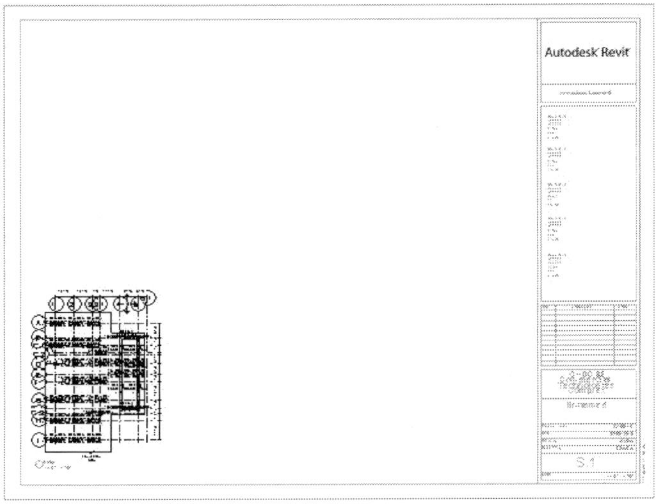

Figure 9-11 Modified location of the ***Entry*** *structural plan view*

10. Next, right-click and choose the **Deactivate View** option from the shortcut menu displayed.

11. Click on the **Camera-view** node under the **3D Views** head in the **Project Browser** and then drag it to the sheet. Next, release the mouse button and click at a location to place the **Camera-view** view, as shown in Figure 9-12.

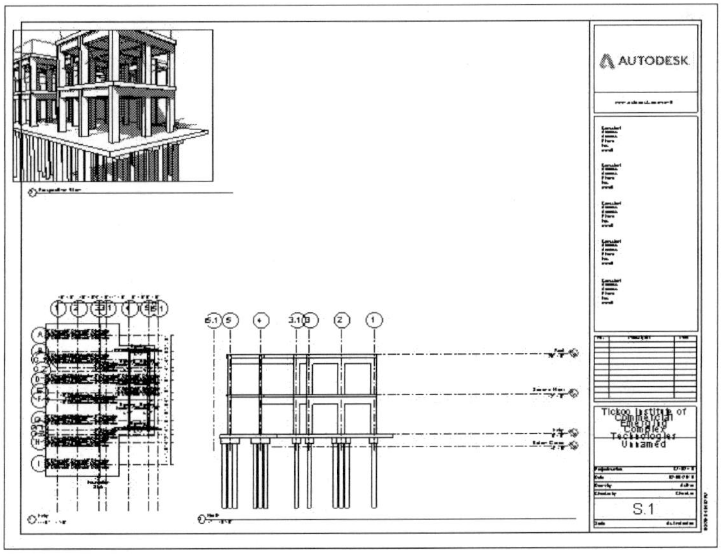

Figure 9-12 Sheet displaying the structural plan, elevation, and camera view

12. Click on the **North** node under the **Elevations (Building Elevation)** head and then specify the value of the **View Scale** parameter to **1" = 10'-0"** for Imperial and **1:100** for Metric. Next, drag the **North** node from the **Project Browser** and place it at the location, as shown in Figure 9-12. The three views are added to the sheet.

Adding Hosted Area Load

In this section, you will add a hosted area load to the structural floors.

1. Double-click on the **Analytical Model** node of the **Project Browser** under the **3D Views** head.

2. Choose the **Visual Style** button in the **View Control Bar** and then choose the **Hidden Line** option from the flyout displayed.

3. Next, choose the **Analyze** tab and then the **Loads** tool from the **Analytical Model** panel; the **Modify | Place Loads** contextual tab is displayed.

4. Choose the **Hosted Area Load** tool from the **Loads** panel. In the **Properties** palette, enter **-.5 ksf** for Imperial and **-23.94 KN/m²** for Metric in the value field of the **Fz 1** parameter and then choose the **Edit Type** button; the **Type Properties** dialog box is displayed.

5. In this dialog box, click on the **Value** field of the **Area force scale** parameter and then enter **10**. Next, choose **Apply** and then **OK**; the **Type Properties** dialog box is closed.

6. Now, in the project view, place the cursor on the edge of the structural floor at the **Roof** level, as shown in Figure 9-13. Next, click on this edge; the assigned load is displayed throughout the area of the floor at the **Roof** level, as shown in Figure 9-14.

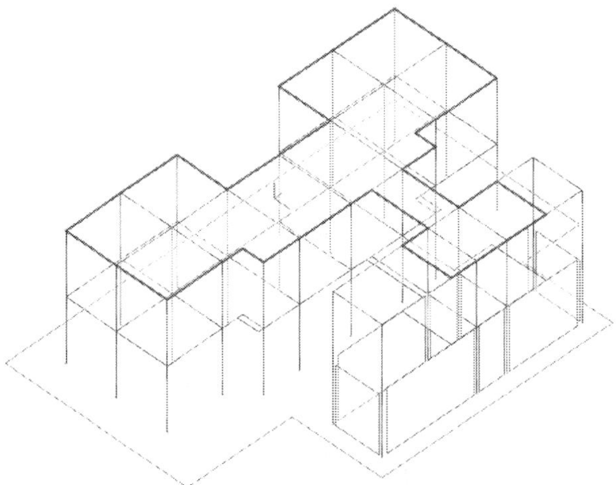

Figure 9-13 *Placing the cursor on the edge of the structural floor*

7. Repeat the procedure followed in step 6 to add the area load to the structural floor on the **Second Floor** level. Press ESC to exit the tool. Figure 9-15 shows the structural model after the hosted area load is added to the structural floor at the **Second Floor** level.

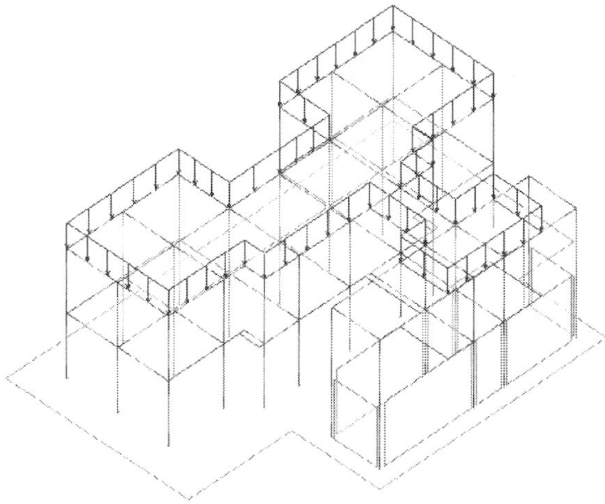

Figure 9-14 *The hosted area load added to the structural floor at the **Roof** level*

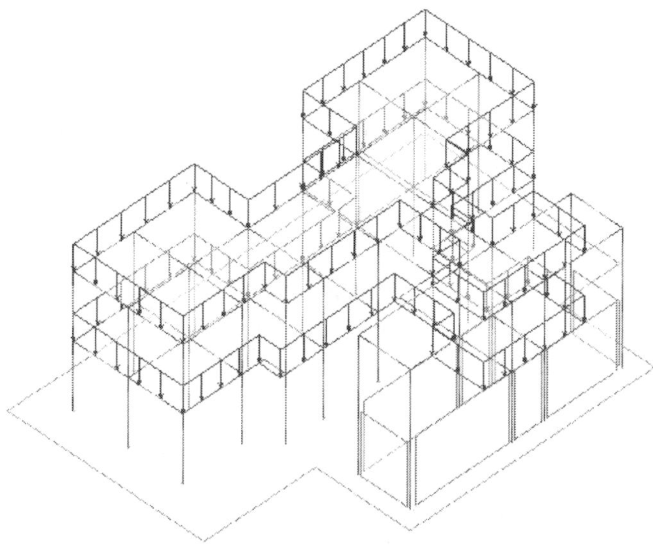

Figure 9-15 *The hosted area load added to structural floor at the **Second Floor** level*

Saving the Project

In this section, you will save the project file using the **Save As** tool.

1. To save the project, choose **Save As > Project** option from the **File** menu; the **Save As** dialog box is displayed.

2. In this dialog box, browse to *C:\rst_2020* and create a folder with the name *c09_rst_2020_tut*. Next, open the created folder and then enter **c09_Commercial-Complex_tut1** for Imperial or **M_c09_Commercial-Complex_tut1** for Metric in the **File name** edit box.

3. Now, choose the **Save** button; the **Save As** dialog box closes and the project file is saved.

Closing the Project

1. To close the project, choose the **Close** option from the **File** menu.

   The file is closed. This completes Tutorial 1 of Chapter 9.

EXERCISE

Exercise 1 Academic Institution

In this exercise, you will create a sheet and add it to the *Academic-Institution* project for Imperial and *M_Academic-Institution* project for Metric, created in Exercise 1 of Chapter 8. For the location of the views in the sheet, refer to Figure 9-16, use the following specifications:

(Expected time: 30 min)

1. Insert the following views in the sheet: **3D View: {3D}**, **Elevation: East**, **Elevation 1-a**, **Elevation 2-a**, and **Structural Plan: Entry**.

2 Use the desired sheet type.
   For Imperial **E130x42Horizontal**
   For Metric **A1 Metric**

3. File name to be referred:
   For Imperial *c08_Academic-Institution_exer1.rvt.*
   For Metric *M_c08_Academic-Institution_exer1.rvt.*

4. File name to be assigned:
   For Imperial *c09_Academic-Institution_exer1.rvt.*
   For Metric *M_c09_Academic-Institution_exer1.rvt.*

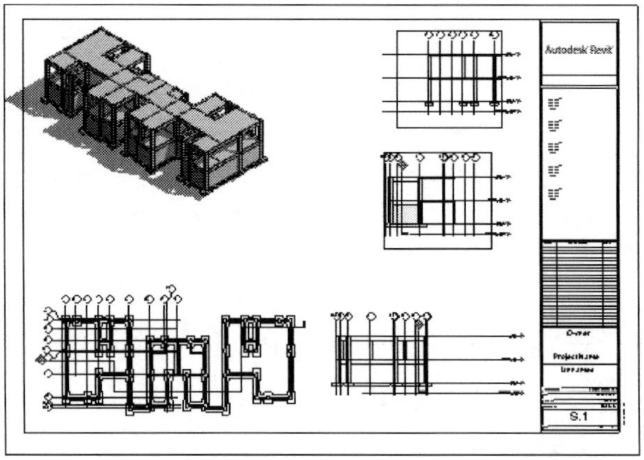

Figure 9-16 *The added views in the sheet*

Chapter 10

Linking Revit Model with Robot Structural Analysis

Learning Objectives

After completing this chapter, you will be able to:

- *Analyze a model using Robot Structural Analysis software*
- *Modify a model in Robot Structural Analysis software*
- *Manage analysis results*
- *Run reinforcement code check*

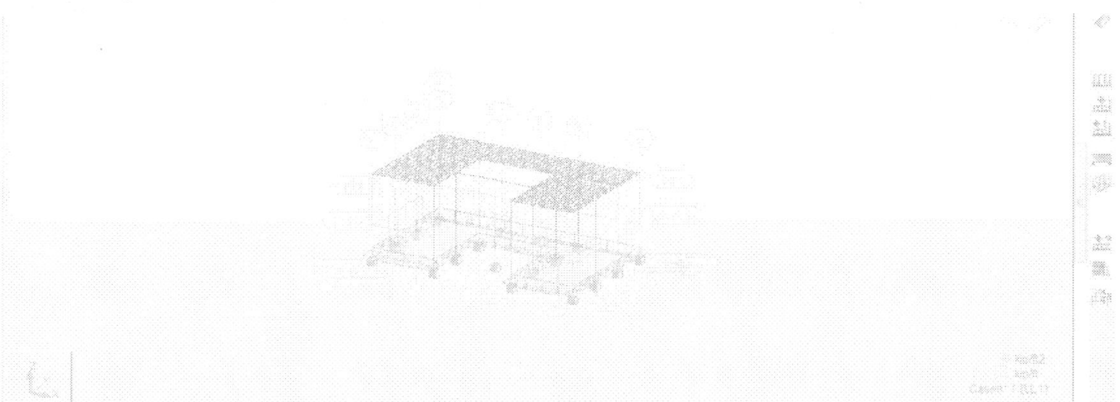

INSTALLING ROBOT STRUCTURAL ANALYSIS PROFESSIONAL 2020

In Autodesk Revit, after developing a model, annotating it, preparing schedules, defining loads, and checking consistency, the model is ready for analysis and code checking. To analyze the model in Robot Structure Analysis 2020, you have to first install the Robot Structural Analysis Professional 2020 software in your system.

You can start Autodesk Robot Structural Analysis by double-clicking on the **Autodesk Robot Structural Analysis Professional 2020** icon on the desktop. Alternatively, choose **Start > Autodesk Robot Structural Analysis Professional 2020 > Autodesk Robot Structural Analysis Professional 2020**. The interface screen of Autodesk Robot Structural Analysis Professional 2020 is shown in Figure 10-1.

STRUCTURAL ANALYSIS INTEROPERABILITY

In Autodesk Revit, you can export the analytical model to Autodesk Robot Analysis and perform structural analysis and also check code on it. You can also perform reinforcement design process of a concrete structure directly by linking it to Autodesk Robot Structural Analysis software. The methods of linking the analytical model from Autodesk Revit 2020 to Autodesk Robot Structural Analysis software are discussed next.

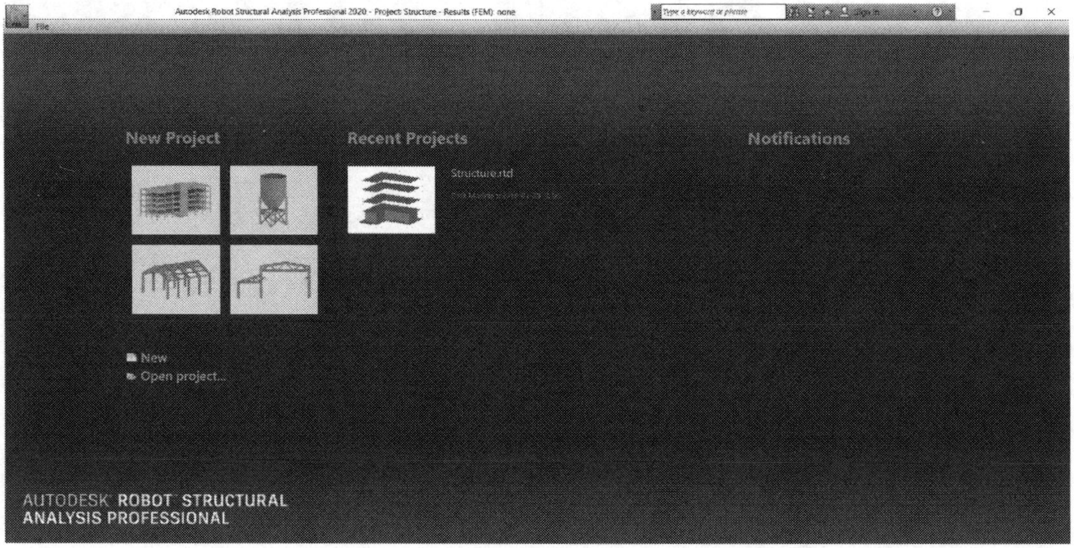

***Figure 10-1** The interface of Autodesk Robot Structural Analysis Professional 2020*

Structural Analysis in Autodesk Robot Structural Analysis Professional 2020

In the previous chapters of this textbook, you learned about various methods of modeling structural elements, imposing loads, setting boundary conditions, viewing reports, setting analytical models, and importing the structural model to a structural analysis software. Now, you will learn how to link Revit files to Autodesk Robot Structural Analysis Professional 2020, perform analysis, and view analysis results in this software. Moreover, you will also learn to update the structural model in Revit based on the analysis and design process performed in Robot.

Autodesk Robot Structural Analysis software, also known as Robot, is a structural analysis and design software that helps you achieve accurate design results for a project. This software provides a collaborative workflow and interoperability with 3D bidirectional link to Autodesk Revit and other Autodesk companion products. The various features and capabilities of Robot Structural Analysis software are discussed next.

TUTORIAL

General instructions for downloading the model file and performing the tutorial:

1. Download the tutorial file **c10_revit_st_2020** from *www.cadsofttech.com*. The path of the file is as follows: *Textbooks > Civil/GIS > Revit Structure > Revit Structure 2020 for Novices*.

2. Next, browse to *C:\rst_2020* and create a new folder with the name **c10_rst_2020_tut**. Next, save and extract the files in this folder.

3. To perform this tutorial, you are required to download and install the **Autodesk Robot Structural Analysis Professional 2020** software.

Tutorial 1 — Building Project

In this tutorial, you will perform the analysis of Revit model in Autodesk Robot Structural Analysis software. **(Expected time: 1 hr 15 min)**

The following steps are required to complete this tutorial:

a. Open the file *Analytical Model.rvt*.
b. Check the model for consistency.
c. Check supports of the model.
d. Send the model to Autodesk Robot Analysis Software.
e. Calculate and prepare results of analysis.
f. Update the Revit model.
g. Saving the project.

Opening the File

In this section, you will open the model.

1. To open the project file, choose **Open > Project** from the **File** menu; the **Open** dialog box is displayed.

2. In this dialog box, browse to *C:\rst_2020\c10_rst_2020_tut* and open the file *Analytical Model.rvt* from this folder.

Checking the Model for Consistency and Supports

In this section, you will check the consistency of the model.

1. Choose the **Analyze** tab and then choose the **Analytical Model Settings** button from the **Analytical Model Tools** panel; the **Structural Settings** dialog box is displayed.

2. In this dialog box, the **Analytical Model Settings** tab is chosen by default. Now, in the **Automatic Checks** area under this tab, select both the **Member Supports** and **Analytical / Physical Model Consistency** check boxes, refer to Figure 10-2.

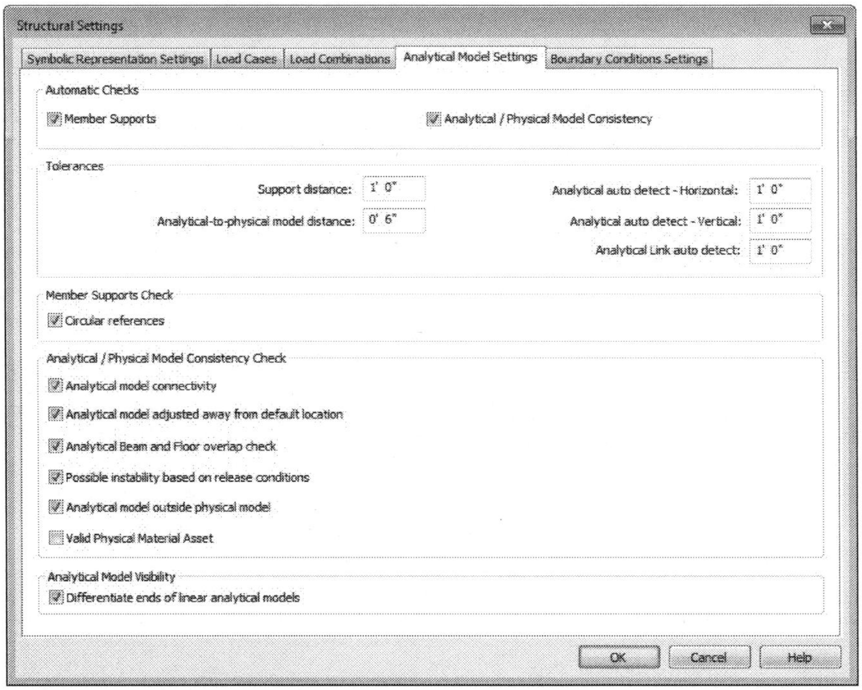

Figure 10-2 *The* ***Structural Settings*** *dialog box*

3. Next, choose the **OK** button in this dialog box; the **Analytical Model Check** message box is displayed, prompting you whether to perform analytical model check or not. Choose the **Yes** button in this message box; the message box and the **Structural Settings** dialog box are closed and a warning message box is displayed.

4. In this message box, choose the **Expand warning dialog** button; the **Autodesk Revit 2020** message box is displayed, as shown in Figure 10-3. View the details of the warnings and choose the **Close** button; the message box is closed. 

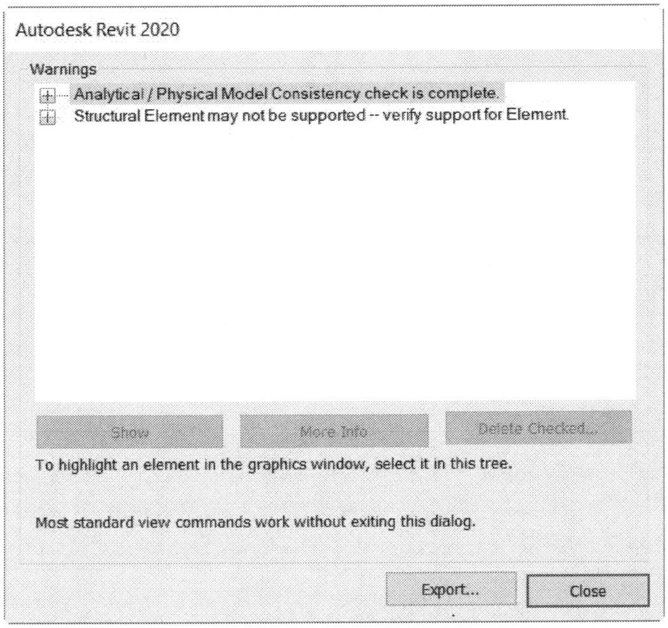

Figure 10-3 *The* ***Autodesk Revit 2020*** *message box*

Sending the Model to Robot Structural Analysis Software

In this section, you will send the model to Autodesk Robot Structural Analysis Professional 2020 software.

1. Choose the **Analytical Model** view under the **3D Views** node in the **Project Browser**, if the current view is different from the **Analytical Model** view.

Note

To view the analytical model, make sure that the ***Show Analytical Model*** *button is chosen from the* ***View Control*** *bar.*

2. Choose the **Robot Structural Analysis Link** tool from **Analyze > Structural Analysis > Robot Structural Analysis** drop-down; the **Integration with Robot Structural Analysis** dialog box is displayed, as shown in Figure 10-4.

3. In this dialog box, ensure that the **Send model** and **Direct integration** radio buttons are selected, and then choose the **Send options** button; the **Integration with Robot Structural Analysis - Send Options** dialog box is displayed, refer to Figure 10-5.

4. In this dialog box, keep the default settings and choose the **OK** button; the **Integration with Robot Structural Analysis - Send Options** dialog box is closed.

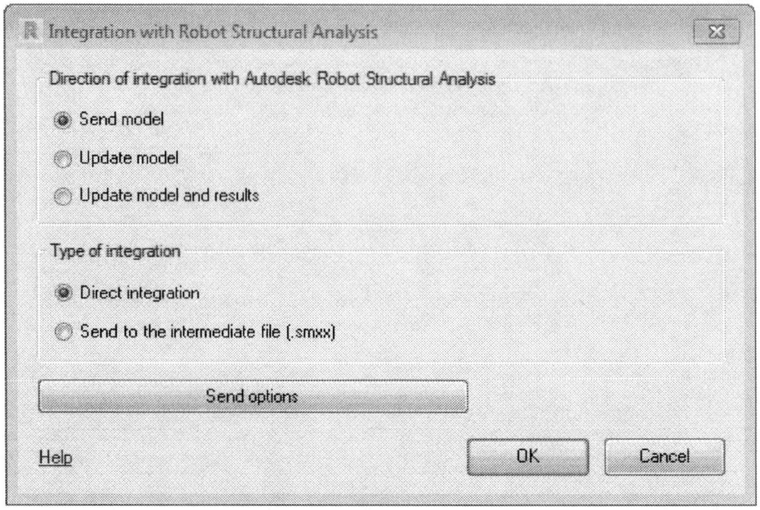

*Figure 10-4 The **Integration with Robot Structural Analysis** dialog box*

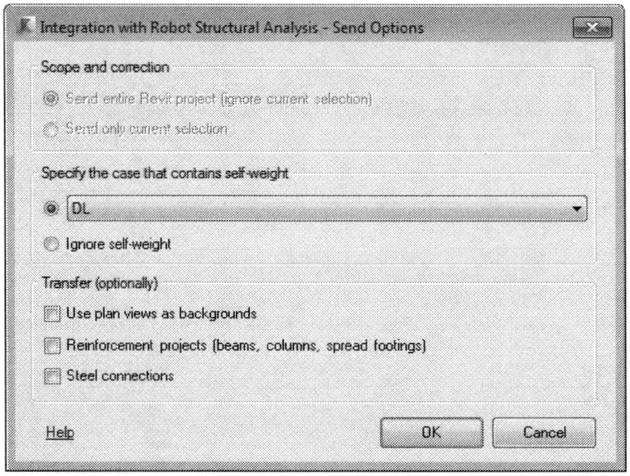

*Figure 10-5 The **Integration with Robot Structural Analysis - Send Options** dialog box*

5. Choose the **OK** button; the **Integration with Robot Structural Analysis** dialog box is closed and **Send model to Robot Structural Analysis** window is displayed. This window displays the status of file transfer and will automatically close on the completion of the file transfer process. Once the file transfer process is complete, the **Integration with Robot Structural Analysis** window is displayed.

6. Choose the **No** button from this window; the window is closed and the model is displayed in the Robot Structural Analysis Professional 2020 application, as shown in Figure 10-6. If the Robot Structural Analysis is already running then the run time error will occur. So make sure you close the Robot Structural Analysis software.

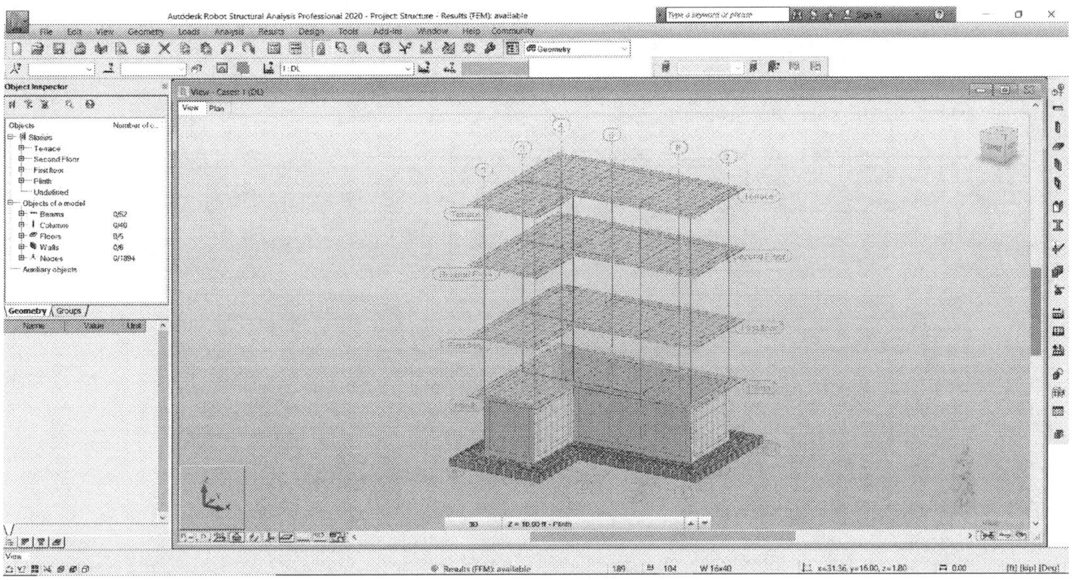

Figure 10-6 *The exported Revit model in the Robot Structural Analysis Professional 2020*

Calculating and Viewing Results

In this section, you will calculate the results in the **Robot Structural Analysis** software.

1. Change the current view of the model to the top view by using the view cube and then select the **Load Types** option from the **Loads** menu; the **Load Types** dialog box with the list of load cases is displayed, refer to Figure 10-7. Choose the **Close** button to close the **Load Types** dialog box.

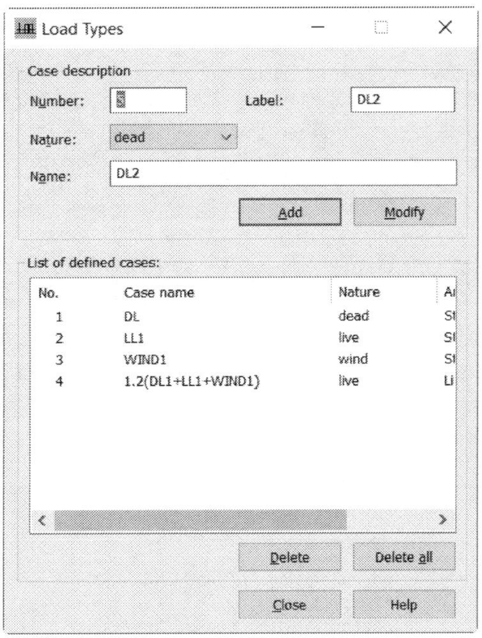

Figure 10-7 *The* ***Load Types*** *dialog box*

2. Choose the **Calculations** tool from the **Analysis** menu, refer to Figure 10-8; the **Autodesk Robot Structural Analysis Professional - Calculations** window is displayed, refer to Figure 10-9. Note that this window displays the calculation status, and is automatically closed when the calculations are completed.

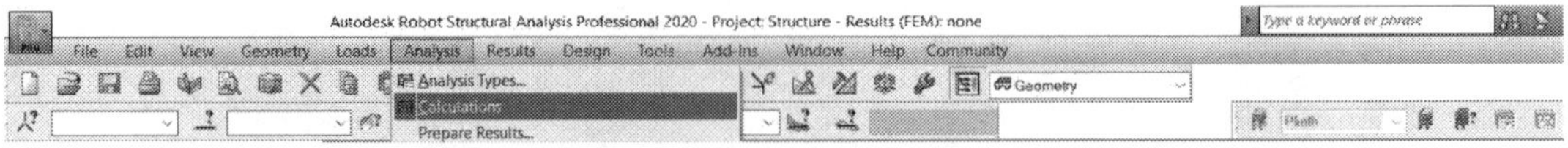

*Figure 10-8 The **Calculations** tool in the **Analysis** menu*

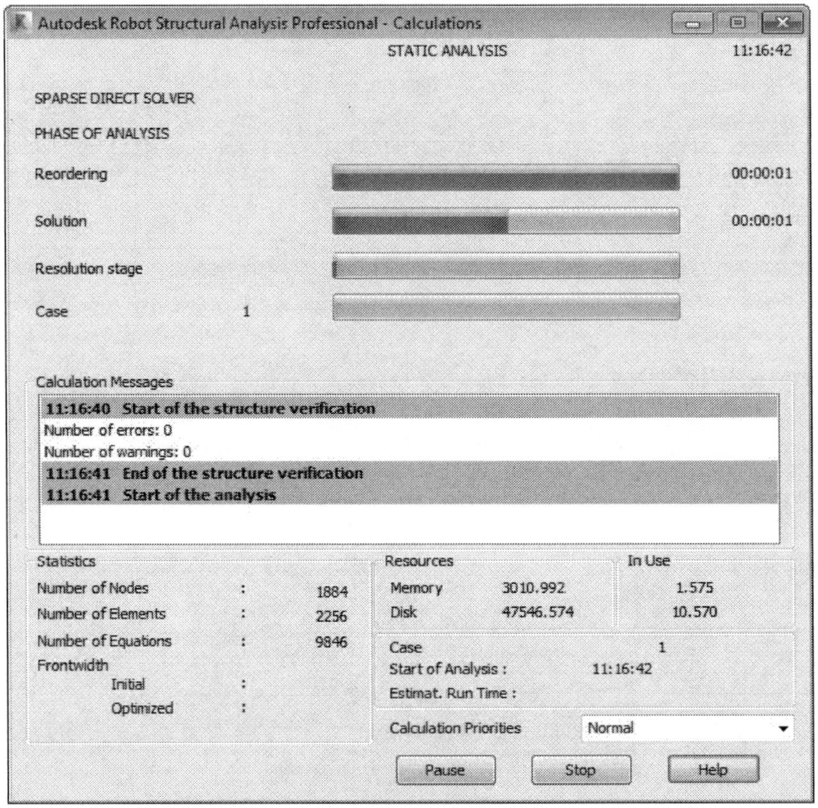

*Figure 10-9 The **Autodesk Robot Structural Analysis Professional - Calculations** window*

3. After performing calculations, choose the **Detailed Analysis** tool from the **Results** menu; the view is changed to **Detailed Analysis** and the **Detailed Analysis** dialog box is displayed.

4. In this dialog box, the **NTM** tab is chosen by default. Select the **Fx**, **Fy**, and **Fz** check boxes in the **Forces** area and the **Mx**, **My**, and **Mz** check boxes in the **Moments** area, refer to Figure 10-10.

5. Next, choose the **Stresses** tab in the **Detailed Analysis** dialog box and select the **maximum - S max** and **minimum - S min** check boxes in the **Normal - maximum** area.

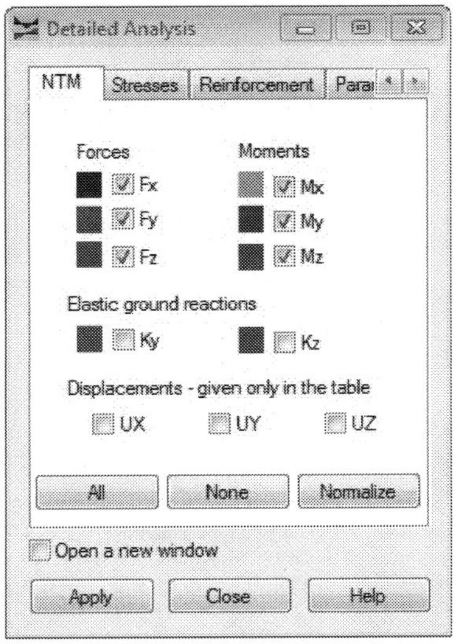

Figure 10-10 *The **Detailed Analysis** dialog box*

6. Choose the **Apply** button; the **Detailed Analysis** window is displayed, showing various graphs for the force and moment analysis, refer to Figure 10-11.

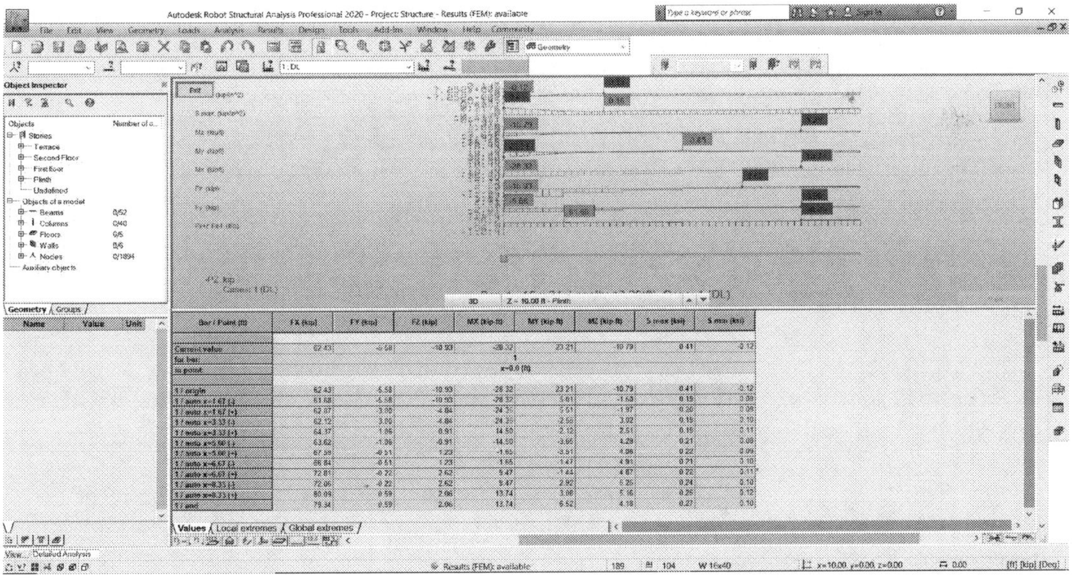

Figure 10-11 *Details of the forces and moments in the changed view*

7. Next, choose the **Close** button in the **Detailed Analysis** dialog box; the dialog box is closed.

Preparing Results

In this section, you will prepare results of analysis.

1. Choose the **Prepare Results** tool from the **Analysis** menu; the **Prepare results** dialog box is displayed, as shown in Figure 10-12. Choose the Browse button located next to the **Template** drop-down; the **Composition of Template for Result Preparation** dialog box is displayed.

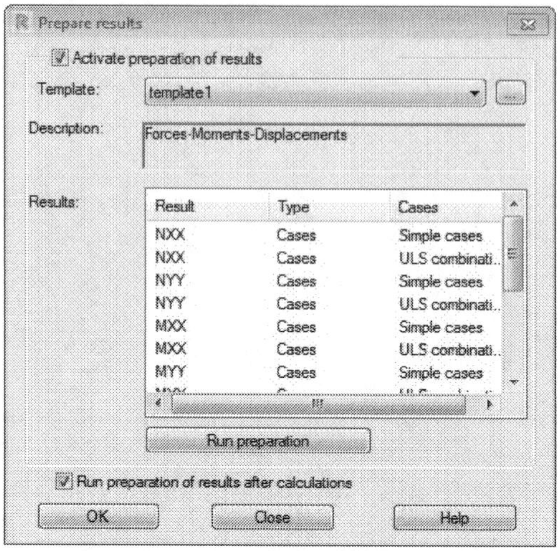

*Figure 10-12 The **Prepare results** dialog box*

2. In the **Template** edit box, enter **Analytical Model** and then choose the **Save** and **Close** buttons; the changes are saved and the **Composition of Template for Result Preparation** dialog box is closed.

3. Choose the **Run preparation** button in the **Prepare results** dialog box and then choose the **OK** button; the **Prepare results** dialog box is closed.

4. Next, choose the **Save** option from the **File** menu of the **Robot Structural Analysis** application; the **Save As** dialog box is displayed. In this dialog box, save the result at an appropriate location.

Updating Revit Model

In this section, you will update the model in Revit.

1. Switch to Revit software and choose the **Robot Structural Analysis Link** tool from **Analyze > Structural Analysis > Robot Structural Analysis** drop-down; the **Integration with Robot Structural Analysis** dialog box is displayed.

2. In this dialog box, select the **Update model and results** and **Direct integration** radio buttons in the **Direction of integration with Autodesk Robot Structural Analysis** and **Type of integration** areas, respectively.

3. Next, choose the **Update options** button in this dialog box; the **Integration with Robot Structural Analysis - Update Options** dialog box is displayed, as shown in Figure 10-13.

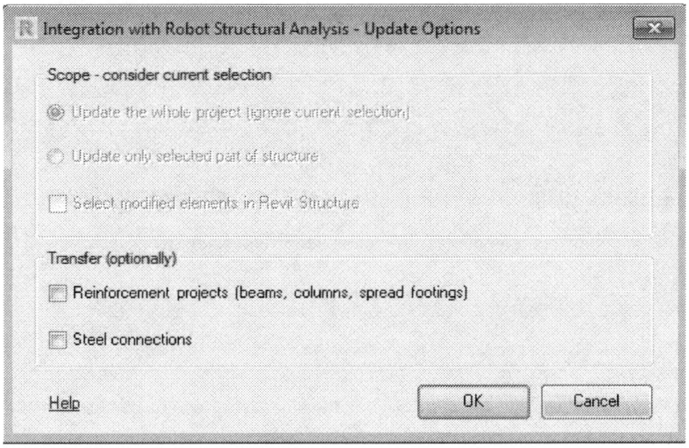

***Figure 10-13** The **Integration with Robot Structural Analysis - Update Options** dialog box*

4. In this dialog box, select the **Reinforcement projects (beams, columns, spread footings)** and **Steel Connections** check boxes in the **Transfer (optionally)** area. Next, choose the **OK** button; the dialog box is closed.

5. Next, in the **Integration with Robot Structural Analysis** dialog box, choose the **OK** button; the dialog box is closed and the **Send Results to Revit** dialog box is displayed.

6. In this dialog box, click in the edit box next to **Model name** in the **Analysis results package** area and enter **Analytical Model**.

7. Enter **Analytical Model** in the **Model name** edit box in the **Required reinforcement results package** area, refer to Figure 10-14.

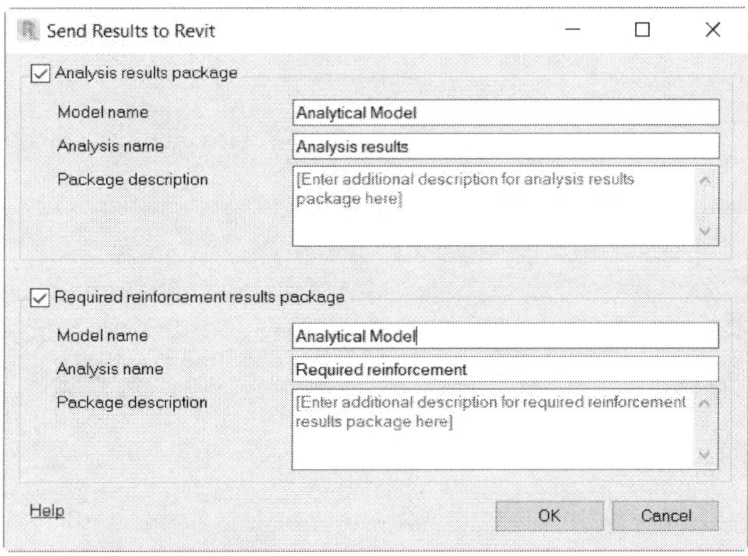

***Figure 10-14** The **Send Results to Revit** dialog box*

8. Next, choose the **OK** button; the **Send Results to Revit** dialog box is closed and the **Integration with Robot Structural Analysis** message box is displayed, as shown in Figure 10-15.

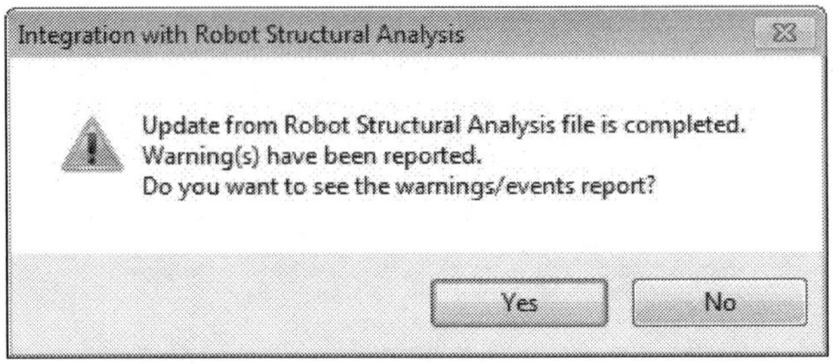

Figure 10-15 *The* ***Integration with Robot Structural Analysis*** *message box*

Note
The process of updating the model will take time and will depend on the processor. The status of the file updation is displayed in the ***Update model from Robot Structural Analysis*** *window.*

9. Choose the **Yes** button from this message box; the **Warning list** message box is displayed, showing update details and warning messages.

10. Choose the **Close** button to close the message box; the model is updated and is saved.

Saving the Project

In this section, you will save the project file using the **Save As** tool.

1. To save the project, choose the **Save As > Project** option from the **File** menu; the **Save As** dialog box is displayed.

2. In this dialog box, browse to *C:\rst_2020\c10_rst_2020_tut* and then enter **Analytical Model_1** in the **File name** edit box.

3. Now, choose the **Save** button; the **Save As** dialog box closes and the project file is saved.

Closing the Project

1. To close the project, choose the **Close** option from the **File** menu.

EXERCISE

Exercise 1 Industrial Shade

Download the Revit file *c10_Industrial_shade_exer* from *https://www.cadsofttech.com*. The path of the file is as follows: *Textbooks > Civil/GIS > Revit Structure > Revit Structure 2020 for Novices*. In this exercise, you will analyze the Industrial Shade file in Robot Structural Analysis software. Refer to Figure 10-16 to view the model in Revit window and Figure 10-17 for the graphical representation of the model in Revit. **(Expected time: 1hr 45min)**

1. Project file to be used: *c10_Industrial_shade_exer.*
2. Link the Revit model to Robot Structure Analysis software.
3. Run calculations.
4. Prepare results.
5. Update the model in Revit.
6. Use the **Result Explorer** tool to view the following results graphically in Revit:
   Rotation of the Reactions
   Displacement in members
   Deformation

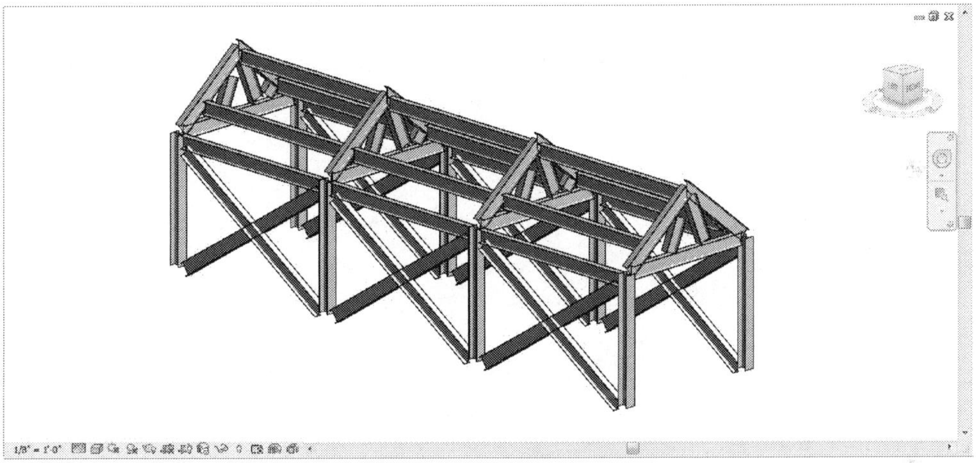

Figure 10-16 *The Steel shade model in Revit window*

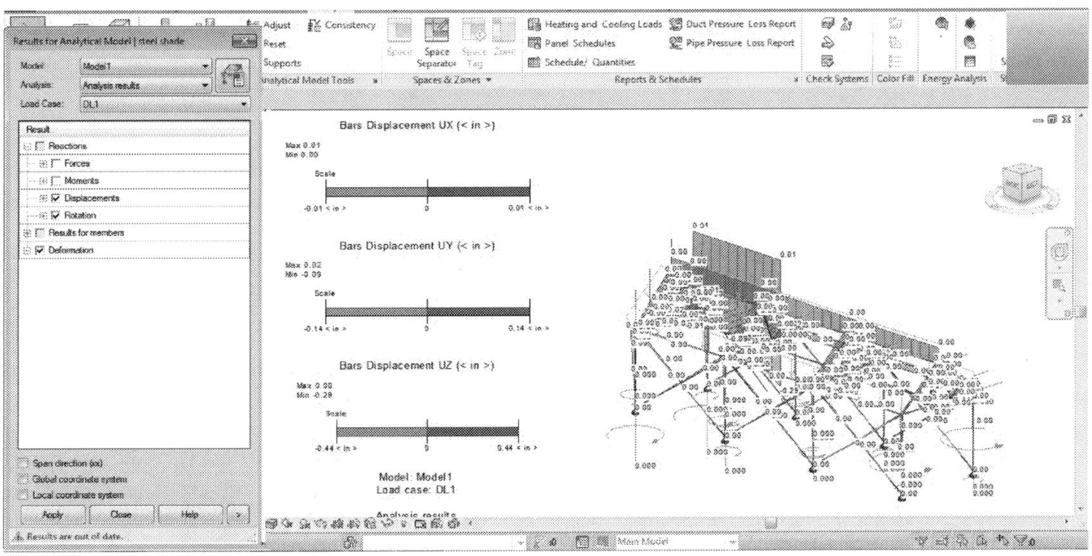

Figure 10-17 *Viewing results by using the* ***Results Explorer*** *tool*

This page is intentionally left blank

Index

P

Q

R

S

T

V

Made in the USA
Monee, IL
13 September 2026

62975274R00098